Intelligent Data Sensing and Processing for Health and Well-being Applications

Intelligent Data Sensing and Processing for Health and Well-being Applications

Edited by

Miguel Wister

Juarez Autonomous University of Tabasco, Cunduacan, Tabasco, Mexico

Pablo Pancardo

Juarez Autonomous University of Tabasco, Cunduacan, Tabasco, Mexico

Francisco Acosta

Juarez Autonomous University of Tabasco, Cunduacan, Tabasco, Mexico

José Adán Hernández

Juarez Autonomous University of Tabasco, Cunduacan, Tabasco, Mexico

Series Editor

Fatos Xhafa

Universitat Politècnica de Catalunya, Spain

ACADEMIC PRESS

An imprint of Elsevier

Academic Press is an imprint of Elsevier
125 London Wall, London EC2Y 5AS, United Kingdom
525 B Street, Suite 1650, San Diego, CA 92101, United States
50 Hampshire Street, 5th Floor, Cambridge, MA 02139, United States
The Boulevard, Langford Lane, Kidlington, Oxford OX5 1GB, United Kingdom

Notices
Knowledge and best practice in this field are constantly changing. As new research and experience broaden our understanding, changes in research methods, professional practices, or medical treatment may become necessary.

Practitioners and researchers must always rely on their own experience and knowledge in evaluating and using any information, methods, compounds, or experiments described herein. In using such information or methods they should be mindful of their own safety and the safety of others, including parties for whom they have a professional responsibility.

To the fullest extent of the law, neither the Publisher nor the authors, contributors, or editors, assume any liability for any injury and/or damage to persons or property as a matter of products liability, negligence or otherwise, or from any use or operation of any methods, products, instructions, or ideas contained in the material herein.

Library of Congress Cataloging-in-Publication Data
A catalog record for this book is available from the Library of Congress

British Library Cataloguing-in-Publication Data
A catalogue record for this book is available from the British Library

ISBN 978-0-12-812130-6

For information on all Academic Press publications
visit our website at https://www.elsevier.com/books-and-journals

Working together
to grow libraries in
developing countries

www.elsevier.com • www.bookaid.org

Publisher: Mara Conner
Acquisition Editor: Sonnini R. Yura
Editorial Project Manager: Ana Claudia A. Garcia
Production Project Manager: Kamesh Ramajogi
Cover Designer: Victoria Pearson

Typeset by SPi Global, India

Contents

Contributors

Francisco Acosta
Juarez Autonomous University of Tabasco, Cunduacan, Tabasco, Mexico

José Adán Hernández
Juarez Autonomous University of Tabasco, Cunduacan, Tabasco, Mexico

Rojas-Dominguez Alfonso
National Technology Mexico/Leon Institute of Technology, Leon, Mexico

Ángel G. Andrade
Universidad Autónoma de Baja California (UABC), Mexicali, Mexico

Aldair Antonio-Aquino
CONACYT, Mexico City, Mexico

Alfonso Avila
Monterrey Institute of Technology and Higher Education, Monterrey, Mexico

Everardo Bárcenas
CONACYT, Mexico City; University of Veracruz, Xalapa, Mexico

Edgard Benítez-Guerrero
University of Veracruz, Xalapa, Mexico

Juana Canul-Reich
Universidad Juárez Autónoma de Tabasco, Cunduacan, Tabasco, Mexico

Lino-Ramirez Carlos
National Technology Mexico/Leon Institute of Technology, Leon, Mexico

Luis A. Castro
Sonora Institute of Technology (ITSON), Ciudad Obregon, Mexico

Raymundo Cornejo
Autonomous University of Chihuahua (UACH), Chihuahua, Mexico

Joaquín Cortez G.
Sonora Institute of Technology (ITSON), Ciudad Obregón, México

Djallel Eddine-Boubiche
University of Batna 2, Batna, Algeria

Adolfo Espinoza I.
Sonora Institute of Technology (ITSON), Ciudad Obregón, México

Doctor Faiyaz
University of Essex, Colchester, United Kingdom

Jesús Favela
CICESE Research Center, Ensenada, Mexico

Armando García B.
Sonora Institute of Technology (ITSON), Ciudad Obregón, México

Muñoz-López Gerardo
National Technology Mexico/Leon Institute of Technology, Leon, Mexico

Enrique Gonzalez
Monterrey Institute of Technology and Higher Education, Monterrey, Mexico

Song Han
University of Connecticut, Storrs, CT, United States

Netzahualcóyotl Hernández
Ulster University, Newtownabbey, United Kingdom; CICESE Research Center, Ensenada, Mexico

Betania Hernández-Ocaña
Universidad Juárez Autónoma de Tabasco, Cunduacan, Tabasco, Mexico

José Hernández-Torruco
Universidad Juárez Autónoma de Tabasco, Cunduacan, Tabasco, Mexico

Faouzi Hidoussi
Corgascience Company, Borehamwood, United Kingdom

Meza-Higuera Jesus A.
National Technology Mexico/Leon Institute of Technology, Leon, Mexico

Kam-Yiu Lam
City University of Hong Kong, Kowloon Tong, Hong Kong

José L. López-Martínez
University of Yucatan-Department of Mathematics, Merida, Mexico

Zamudio-Rodriguez V. Manuel
National Technology Mexico/Leon Institute of Technology, Leon, Mexico

Fernando Martínez
Autonomous University of Chihuahua (UACH), Chihuahua, Mexico

Carmen Mezura-Godoy
University of Veracruz, Xalapa, Mexico

Guillermo Molero-Castillo
CONACYT, Mexico City; University of Veracruz, Xalapa, Mexico

Julio Muñoz-Benítez
University of Veracruz, Xalapa, Mexico

David Munoz-Rodriguez
Monterrey Institute of Technology and Higher Education, Monterrey, Mexico

Joseph Kee-Yin Ng
Baptist University of Hong Kong, Kowloon Tong, Hong Kong

Pablo Pancardo
Juarez Autonomous University of Tabasco, Cunduacan, Tabasco, Mexico

Ioannis Papavasileiou
University of Connecticut, Storrs, CT, United States

Raul Peña
Monterrey Institute of Technology and Higher Education, Monterrey, Mexico

Umair M. Qureshi
City University of Hong Kong, Kowloon Tong, Hong Kong

Marcela D. Rodríguez
Universidad Autónoma de Baja California (UABC), Mexicali, Mexico

Luis-Felipe Rodríguez
Sonora Institute of Technology (ITSON), Ciudad Obregon, Mexico

Baltazar-Flores Rosario
National Technology Mexico/Leon Institute of Technology, Leon, Mexico

Erica Ruiz I.
Sonora Institute of Technology (ITSON), Ciudad Obregón, México

Gabriela Sánchez
CONACYT, Mexico City; University of Veracruz, Xalapa, Mexico

Jan Sliwa
Bern University of Applied Sciences, Biel, Switzerland

Homero Toral-Cruz
Center for Research in Mathematics, University of Quintana Roo, Chetumal, Mexico

Joel A. Trejo-Sánchez
CONACYT-Center for Research in Mathematics, Merida, Mexico

Nelson Wai-Hung Tsang
City University of Hong Kong, Kowloon Tong, Hong Kong

Ochoa-López Verónica del Rocio
National Technology Mexico/Leon Institute of Technology, Leon, Mexico

Zamudio-Rodríguez Victor
National Technological Institute of Mexico/Technological Institute of León, Mexico

Miguel Wister
Juarez Autonomous University of Tabasco, Cunduacan, Tabasco, Mexico

Antonio Xohua-Chacón
University of Veracruz, Xalapa, Mexico

Foreword

We currently observe rapid development in the area of sensing properties of the physical world and in processing the collected data. Due to progress in sensor technology, we can now measure many properties that were not easily available before. Miniaturization of these devices permits access to difficult locations in a noninvasive way. Low power consumption and energy harvesting make long operation without service possible. Lower prices and mass production permit pervasive deployment.

An important application field for these new technologies is supporting health and well-being. Many vital parameters can be measured. The (incomplete) list starts with classical items such as pulse and blood pressure and goes on to chemical properties (glucose level, substance concentration) and electrical values (electrocardiogram), to movement parameters measured with accelerometers. Sensing can result in action, as with glucose pumps or heart defibrillators. Data streams can be analyzed locally and presented to the user/patient.

This massive data collection calls for reasonable storing, processing, and analysis. Data from individuals can help them to monitor their own health and fitness states and to take appropriate actions, like exercising more or changing nutritional habits. Collecting global data permits monitoring of population health and enhancing medical knowledge. However, such data are extremely sensitive and have to be adequately protected. Even an apparently harmless application such as counting steps with a smartphone permits location of each user. In the enthusiasm for Big Data, such issues cannot be forgotten.

This book covers various aspects of sensing in the area of health and well-being. It consists of three parts. The first part presents the basics of current smart sensor technology and data processing architectures. The second part treats the application of sensing in health and well-being. Special stress is put on ambient assisted living environments and smart homes. In addition, security, privacy, and ethical issues are discussed. The third part presents various specific examples of such applications. Several chapters treat the mining and analysis of collected data to improve diagnosis and treatment of certain medical conditions.

This book should be helpful for people wanting to enter this new and important field. The reader will get a current overview of the available technology as well as a collection of practical examples that can provide inspiration for future developments.

Jan Sliwa
Bern University of Applied Sciences, Bern, Switzerland

Preface

Increasingly in our day-to-day life, we humans encounter sensors being used to collect data from the environment. Through special treatment of this data, we are kept better informed of what is happening around us. More recently, we are encountering devices that sense the human body itself, detecting changes that occur in our bodies related to vital signs or diseases.

It seems that every day more devices appear, such as mobile phones and wireless sensors with communication capabilities, that are placed on our bodies and in the near environment, with the ultimate purpose of improving our quality of life. Mobile devices collect data to be analyzed using computational tools that give results using some degree of intelligence.

This book explores the fundamental issues related to the current use of sensing devices, intelligent data acquisition, and processing, as well as applications and information, with a focus on health and well-being applications.

The central focus of the book is application-oriented and is aimed at the area of datacentric systems/intelligent data sensing. There are many potential applications and many proposals and prototypes have appeared in this area, but at this point only a few matured applications have been implemented in real life. This is likely to change because of the recent introduction of new sensor technologies and the Internet of Things (IoT), and their application to real-life problem solutions. Therefore, this book reviews the fundamental concepts of gathering, processing, and analyzing data from devices disseminated in the environment, as well as the latest developments in collection, processing, and abstraction of datasets and smart mobile data sensing. All of these phases represent a natural evolution of ubiquitous computing, aiming towards the IoT. The goal is to be present to the internet and to connect any useful device for users, to obtain added value. The book covers sensor-collected and processing intelligence for health and well-being applications, dealing not only with technical issues but also with issues involving compliance with security, privacy, and ethical standards in smart sensor applications for health and well-being.

- It introduces concepts and emerging techniques and technologies needed to understand sensor-collected intelligence for health care and well-being in the workplace and at home. The concepts described in these pages have the potential to offer realistic views from an application perspective and to reveal real-life issues in design, development, deployment, etc.
- It reviews recent works related to the current use of sensing devices, intelligent acquisition, and processing, as well as applications and first-hand information that the authors have developed.
- It discusses the latest views on security, privacy, and ethics in smart sensor applications for health and well-being applications.

The book is organized into three parts, comprising 14 chapters. We briefly introduce each chapter here.

Part 1. Introduction to Smart Sensors

Chapter 1. Charting the past, present, and future in mobile sensing research and development.

Luis A. Castro, Marcela D. Rodríguez, Fernando Martínez, Luis-Felipe Rodriguez, Ángel G. Andrade, and Raymundo Cornejo.

In this chapter, the authors present an analysis of different trends that have been extending the mobile sensing field throughout its development, particularly in the area of health care and wellness. The authors have carried out a review of several related papers using a variety of criteria.

Chapter 2. Data fusion architecture of heterogeneous sources obtained from a smart desk.

Julio Muñoz-Benítez, Guillermo Molero-Castillo, and Edgard Benítez-Guerrero.

The subject of this chapter is obtaining homogenized data to be used in analysis and in inferences based on a specific context; the authors propose a conceptual design of a data fusion architecture for the extraction, preprocessing, fusion, and load processing from diverse data sources. Later in the chapter, methods are implemented for extraction, preprocessing, and data fusion according to their nature, to homogenize them and to maintain the coherence of the data.

Chapter 3. Wireless sensor technology for intelligent data sensing: research trends and challenges.

Djallel Eddine-Boubiche, Joel A. Trejo-Sánchez, Homero Toral-Cruz, José L. López-Martínez, and Faouzi Hidoussi.

Some data aggregation methods to optimize the data collection process and sensor nodes are described in this chapter. The authors explain some interesting routing protocols that optimize communication among nodes in a wireless sensor network (WSN) and some strategies for sensor node mobility.

Part 2. Sensing in Health and Well-Being Applications

Chapter 4. Tangible user interfaces for ambient assisted working.

Antonio Xohua-Chacón, Edgard Benítez-Guerrero, and Carmen Mezura-Godoy.

This chapter relates to ambient assisted working (AAW), as the authors explain the characteristics of tangible user interfaces (TUIs), addressing their advantages as well as their limitations. The integration of TUIs into AAW is proposed as an alternative for interaction between users and interactive systems present in a work environment.

Chapter 5. Ambient assisted working applications.

Pablo Pancardo, Miguel Wister, Francisco Acosta and José Adán Hernández.

An architectonic design for ambient assisted working systems, considering users and their context to offer customized results, is proposed in this chapter. This chapter includes an ambient assisted working method to capture and process user profile and context data to deliver customized results to users. Two applications are illustrated: estimation of heat stress in workplaces (HSW) and classification of perceived exertion.

Chapter 6. Smart home automation architecture for comfort, security, and resource savings.

Armando G. Berumen, Erica R. Ibarra, Joaquín C. González, and Adolfo E. Ruiz.

An architecture is presented for home automation that is intended for modern houses. The architecture comprises a coordinator node and several remote nodes communicating through ZigBee. Each remote node is responsible for controlling different activities. A friendly interface is used to communicate with the end user.

Chapter 7. Security, privacy, and ethical issues in smart sensor health and well-being applications.

Jan Sliwa.

The author takes a broad perspective and discusses security, privacy, and ethical issues regarding sensor-based smart medical devices. The author makes the case that these devices offer new opportunities and also create new risks. As shown, risks may be caused by one's own poor design, or by the malicious actions of others.

Chapter 8. Diagnosing medical conditions using rule-based classifiers.

Juana Canul-Reich, Betania Hernandez-Ocaña, and José Hernández-Torruco.

Diagnostic models are some of the resulting applications of data mining in the field of medicine. In this chapter, the authors use four publicly available datasets to create rule-based diagnosis models.

Learning methods applied include JRip, OneR, and PART. More complex learning methods such as SVM and kNN are also used. The idea is to compare all the results and eventually derive conclusions as to the performance of simple rule-based models against these latter techniques. Results show rule-based classifiers are comparable in performance to the more complex models.

Part 3. Smart Sensor Applications for Health and Well-Being

Chapter 9. Assessing the perception of physical fatigue using mobile sensing.

Netzahualcóyotl Hernández and Jesús Favela.

Fatigue assessment is often performed through clinical studies, physical tests, or self-report. Here the authors present a method for assessing an individual's perception of physical fatigue while walking. The approach is based on accelerometer and location data collected from smartphones, thus allowing for the continuous assessment of the perception of physical fatigue under naturalistic conditions.

Chapter 10. Applications to improve the assistance of first aiders in outdoor scenarios.

Enrique Gonzalez Guerrero, Raul Peña, Alfonso Ávila, and David Muñoz.

This chapter presents a review of the m-Health systems for tracking patients' health in outdoor scenarios. Systems are ranked according to their potential for improving remote consultations in real time. After this review, the authors propose an integrated and dynamic system (architecture) able to monitor a patient's physiological parameters. They also discuss technical challenges and current boundaries related to the m-Health concept, based on the proposed system. Finally, they present conclusions regarding the advantages and limits of m-Health systems.

Chapter 11. Indoor activity tracking for elderly using intelligent sensors.

Nelson W.-H. Tsang, Kam-Yiu Lam, Umair M. Qureshi, Joseph K.-Y. Ng, Ioannis Papavasileiou, and Song Han.

In this chapter, the authors discuss how to apply the latest intelligent sensor technologies to track common indoor activities performed by elderly persons in their living rooms, and to detect falls. The authors present two systems: SmartMind (3D camera based), applied for effective activity tracking of the user within a predefined environment, and ActiveLife, in which simple motion sensors are adapted to measure the changes in motion for indoor activity estimation.

Chapter 12. User-centered data mining tool for survival-mortality classification of breast cancer in Mexican-origin women.

Guillermo Molero-Castillo, Everardo Bárcenas, Gabriela Sánchez, and Aldair Antonio-Aquino.

This chapter proposes a classification system for user-centered analysis. This system studies the survival-mortality rate in Mexican-origin women diagnosed with breast cancer. The system is based on a methodology of user-centered data mining, which has as its foundation the principles of the ISO 9241:210:2010 standard. The system is composed of two classification algorithms: logistic regression and support vector machine.

Chapter 13. Modeling independence and security in Alzheimer's patients using fuzzy logic.

Jesus A. Meza-Higuera, Victor M. Zamudio-Rodriguez, Faiyaz Doctor, Rosario Baltazar-Flores, Carlos Lino-Ramirez, Alfonso Rojas-Dominguez.

This chapter contains a model based on fuzzy logic that allows a balance between independence and security in an intelligent environment for monitoring and care of people with Alzheimer's disease. This model reacts according to the initial data given and received from the environment throughout the process.

Chapter 14. Wireless sensor networks applications for monitoring environmental variables using evolutionary algorithms.

Carlos Lino-Ramirez, Víctor M. Zamudio-Rodríguez, Verónica R. Ochoa-López, and Gerardo Muñoz-López.

This chapter mainly deals with the implementation of two types of applications for monitoring environmental variables, one of them focused on an irrigation system and the other one on monitoring air quality. The first application concerns a WSN for which a model is proposed that allows the improvement of irrigation in greenhouses of plants of different species and uses water efficiently. The model proposes the use of metaheuristic algorithms in order to obtain an optimal solution of water consumption. The second proposal is a wireless monitoring system for air pollution, creating a scenario model, simulation, and analysis.

Miguel Wister, Pablo Pancardo

INTRODUCTION TO SMART SENSORS

CHARTING THE PAST, PRESENT, AND FUTURE IN MOBILE SENSING RESEARCH AND DEVELOPMENT

Luis A. Castro*, Marcela D. Rodríguez†, Fernando Martínez‡, Luis-Felipe Rodríguez*, Ángel G. Andrade†, Raymundo Cornejo‡

Sonora Institute of Technology (ITSON), Ciudad Obregon, Mexico* Universidad Autónoma de Baja California (UABC), Mexicali, Mexico† Autonomous University of Chihuahua (UACH), Chihuahua, Mexico‡

1.1 INTRODUCTION

With the advent of smartphones and mobile technologies capable of sensing the environment at reasonable costs, an emerging area has been helping researchers capture data from large groups of populations scattered across large regions. This area, dubbed mobile sensing, has been gaining traction as research relying on mobile technologies has been increasingly carried out in the wild, yielding better results with ecological validity. Mobile sensing allows researchers to collect data at precise times and locations, and these sensors are stored in remote repositories for detailed scrutiny. For instance, data can be collected through sensors (e.g., GPS, accelerometer, and microphone) used by older adults to infer their functional status (e.g., frailty syndrome, mobility, and physical activity) and compared to data obtained by physicians during assessment interviews. However, utilizing mobile phones for research purposes does not bind the findings just to an individual's affairs, but allows the scope of research to extend far beyond the immediate proximity of the phone or sensor, into the surrounding environment, whether physical or societal. For example, mobile phones can be used to map social behavior that can be linked to reported levels of wellbeing within a city.

The field of mobile sensing will be instrumental in several areas of research and development. In particular, research in health care is poised to be infused with mobile sensors derived from the precision medicine initiative (PMI) [1]. In this medicine model, medical treatments are tailored based on a personalized approach, taking into account individual differences (e.g., genetic, contextual, environmental, and behavioral). In this approach, intelligent datacentric computerized systems can be useful for defining better medical treatment and health outcomes.

The use of mobile technologies for wellness—mobile phones in particular—has been increasing over the last few years. This is particularly true for mobile phone applications that aim at increasing a person's wellness, such as calorie counter, physical activity, and socialization apps. This also holds true for the development of commercial sensors such as Fitbit (http://www.fitbit.com) or much more specialized brands such as Polar (http://www.polar.com), which are usually accompanied by mobile phone applications to notify findings.

Intelligent Data Sensing and Processing for Health and Well-being Applications. https://doi.org/10.1016/B978-0-12-812130-6.00001-9

In this chapter, we present an analysis of the pervasive trends in the mobile sensing field. We first identify relevant studies by following a research procedure for conducting scoping reviews, which includes a search strategy to identify published studies, defining an inclusion and exclusion criteria, papers selection, abstracting and charting relevant data, and summarizing and reporting the results. A scoping review is defined as "a type of literature that identifies and maps the available research on a broad topic" [2], which can be used for varied purposes [3]. Then, we analyze the selected studies to identify how technology has provided a large range of types of sensors and how researchers use them. Other papers have reported similar studies [4–17], but to the best of our knowledge, this is the first work aimed at presenting such analysis in the area of health care and wellness. We have focused our analysis on understanding what types of studies have been carried out using mobile sensing that have an impact on health care and wellness.

In particular, we reviewed the methodological design of each paper to identify data that enabled us to determine the type of contribution and to assess the quality of the evidence provided. With this information, we proceeded to classify each paper according to a set of criteria defined by the authors of this paper, such as the sensing paradigm utilized, the subject area of the publication, the types of sensors used in the study, and the application domain to which the studies were applied, mainly health care and wellness.

As a result, we characterize the evolution of the area mainly in terms of the types of sensors used in mobile sensing studies (see Section 1.2). Since 2003, this research area has been gradually developing and can be split into three major research interests that have been gaining attention as the area unfolds: (1) the construction of custom sensing devices, (2) the use of on-device built-in sensors, and (3) the use of commercial sensors. As shown in the Results section, most studies follow the opportunistic sensing paradigm rather than the participatory sensing paradigm. In terms of sensors, researchers have been increasingly using external sensors, as many of them are available off-the-shelf and provide reliable data. Also, during the first years, much of the efforts in the field of mobile sensing were aimed at constructing new sensors. Lately, researchers have been concerned with making sense of the data collected.

In terms of health care and wellness, we found that extensive research has been carried out in engineering areas. Although some studies have included researchers from the field of medicine, they represent only a small proportion of the total number of papers.

To present this review, we organize the book chapter as follows: (i) we first introduce the three main research interests in mobile sensing development; (ii) then, the methodology used for searching the studies analyzed in this work is described; (iii) next, the aggregated results are described; and (iv) finally, we discuss the future of the field of mobile sensing.

1.1.1 RESEARCH AND DEVELOPMENT IN MOBILE SENSING

Mobile phone sensing is a relatively new area of research that has come into existence in part due to the development of tiny, cheap sensors that can be incorporated into mobile devices. This research area, as such, has been gradually developing during the last 15 years and can be split into three different focuses of attention that have been gaining momentum as the area unfolds: (1) the use of custom sensing devices, (2) the use of on-device built-in sensors, and (3) the use of commercial sensors. These three research divisions speak for the development of sensors in particular and the manner in which this development has been influencing the area. As shown in this paper, the development of sensors for mobile sensing was initially led by researchers in academia and later by those in industry.

1.1.1.1 The use of custom sensing devices

The first major research focus in mobile sensing used several devices that were built to meet a particular research need. These sensing devices were mainly crafted after the technology was mature enough to combine sensors that would fit in a box that subjects could carry with them in an unobtrusive way. In some studies, sensing devices with storage mechanisms were used and researchers analyzed the information after downloading it from these devices [18–20]. An early example of a custom sensing device is the Sociometer [20], a shoulder-mounted device with an embedded accelerometer, an infrared (IR) sensor, and a microphone.

As the capabilities of mobile phones increased, they were often used along with custom sensing devices. Advanced feature phones could store a considerable amount of information and communicate with external sensors and remote servers, and also had acceptable processing power. While the custom-made device sensed the environment (or the subject), the phone usually communicated with this device via Bluetooth to collect and save the data in its storage or send it through the network to remote servers for analysis. Also, advanced feature phones enabled researchers to inform subjects of their status as well as correct bad sensor readings. One such example can be found in Ref. [21], where the user received feedback on her physical activities with a glance at the screen of the phone.

Nowadays, modern mobile phones or smartphones include several kinds of sensors. In addition, many specialized sensors are on the market. However, many of those sensors are not suitable for all types of studies. Researchers will always find new variables to measure using different means, so custom-built sensing devices will still be needed, for the time being. As sensors become more manageable and smaller, they could be easily concealed (see Ref. [22]) or embedded.

This first research focus of mobile sensing, although limited, was important as it introduced new ways to sample the outer world, introducing different techniques for observation and data collection.

1.1.1.2 The use of built-in sensors

The second field that gained momentum in mobile sensing research started with the emergence of smartphones. By 2007, mobile phones began to embed sensors for a better user experience (e.g., accelerometers and gyroscopes) and for novel types of services that involved knowing the user's location and orientation (e.g., GPS and compass). With these augmented mobile devices, researchers began to exploit the advantages of their ubiquity and pervasiveness, in addition to their increased capabilities of perceiving and measuring the outer world and their ever-increasing storage, processing, and communication capabilities.

Recent papers have shown that researchers are able to infer several aspects of subjects, like the quality of their sleep [23,24], their level of stress [25], their wellbeing [24], their surroundings [26,27], and even personality traits [28,29]. The usefulness of mobile phones with built-in sensors does not end at the personal level; they also have contributed at the social level, where researchers look for ways to infer social behavior and interaction patterns [30–33], or at the community level, where they help map and identify urban situations and tendencies, like identifying noise pollution in a city [34], mapping potholes, bumps, and chaotic places in a city [35], or predicting bus arrival [36].

Even though the development of built-in sensors in mobile phones and other wearable devices is improving, being able to infer information on situations in the outer world is still an open problem. This is mainly due to the changing nature of the context being inferred, apart from the fact that sensor readings are often noisy. Some real-world situations exacerbate that problem: for instance, carrying a mobile phone in the pocket or purse.

As new sensors are embedded in mobile phones, the number of ways to measure the outer world will increase. At the same time, finding alternate means of using certain sensors will also unfold. For instance, utilizing Bluetooth or microphones as social sensors has become very useful when working with groups of people.

While this research focus area has certainly gained momentum, and mobile phones indeed have many capabilities, certain studies demand on-body sensors placed at particular locations for increased accuracy. For instance, a heart rate monitor at the wrist can enable continuous readings rather than the sparse, often clumsy, readings obtained from a similar sensor on a mobile phone.

1.1.1.3 The use of commercial sensors

As sensors became embeddable and the market matured, devices were created to measure several variables. Health-related gadgets that include sensors such as Baumanometers, heart-rate monitors, pedometers, or calorie counters have been on the market for some years. With the arrival of smartphones and standards for personal area networks, a broad set of applications for these devices began to arise.

Commercial sensors not only help users in their daily lives, but they have also been adopted by the research community for different purposes. In the literature, two main uses for commercial sensors along with mobile phones were found: (1) as a tool to get the ground truth (i.e., they are used to compare with the results of a study using an alternate method), and (2) as a companion to the mobile phone. To illustrate the former, in Ref. [23] they developed an algorithm that uses built-in mobile phone sensors to infer when a person is sleeping and the amount of time they slept. Then, they compared the results with commercial gadgets that compute that information: Jawbone's UP that uses a wrist band, and the Zeo Sleep Manager Pro that uses a head band. To illustrate the latter, in Ref. [37] researchers used the Emotiv Epoc Electroencephalography (EEG) headset to control the address book dialing app of the smartphone with the mind of the subject.

Generally, the usage of external commercial sensors seems to be influenced by the limits of technological development in mobile phones. That is, the device placement and sensors available make the mobile phone suitable for some studies but very limited for others. However, often in these types of studies mobile phones are used to store, process, and send data to remote servers as well as to offer feedback to users.

1.2 METHODS

The scoping review was conducted by the authors of this paper, who have expertise in the development of platforms for carrying out sensing campaigns and in the development of medical informatics technologies. The authors worked in pairs to conduct the search, selection, data abstraction, and analysis of three randomly assigned groups of papers. Thus, each pair read a subset of all the studies (title and abstract for screening, and full text for final selection) independently. To increase reliability they discussed ambiguities of inclusion criteria until consensus was reached. At the end of each methodology phase, results were presented to the overall team to be discussed.

1.2.1 SEARCH AND SELECTION PROCESS

To identify high-quality literature to be included in this paper, we conducted the following search process:

1. Search on Google Scholar. We used the search phrase "mobile sensing" and retrieved the references from the first five results pages. Our search was conducted in December 2016.

2. "Snowballing" [37]. We scanned the reference lists from the full text papers that matched the inclusion criteria, and selected the ones that included the following keywords: mobile, phone, smartphone and sensing in their title, or a composite phrase using two or more of those words. The selected references were looked up on Google Scholar to filter out the ones that were not relevant for the aim of this review.

We used the following criteria to select relevant studies:

- Papers published in journals, conferences, and book chapters.
- Papers written in English.
- Papers reporting results that contribute to the development of the mobile sensing area and/or that apply mobile sensing technologies to study a research problem, e.g., a health-care domain problem.

These criteria were used in a peer-review manner by each subteam. A first selection stage was based on screening and binary rating of abstracts (0: exclude; 1: consider for inclusion). To facilitate the management of study references, they were uploaded to Mendeley reference managing software, which was also used to discard duplicate papers. Afterwards, a review and assessment were carried out on full papers to extract data from the research reported.

1.2.2 DATA ABSTRACTION

Each paper was assigned to two reviewers (authors of this paper), who completed an online form with each paper's descriptive information such as publication year and type (journal, conference, or book chapter). Additionally, reviewers agreed on how to categorize papers based on a taxonomy of study characteristics that was defined to determine the addressed issues, open challenges, and limitations and opportunities for future research.

1.2.2.1 Reviewed papers' contribution

We reviewed the methodological design of each paper to extract data that enabled us to determine the type of contribution and to assess the quality of the evidence provided:

- *Study type*: We identified the main contribution of each paper:
  o A survey/review to map the relevant literature on mobile sensing.
  o A framework or toolkit designed for addressing challenges associated with the development of sensing campaigns.
  o A sensing study conducted to understand the behavior of some variables, which includes identifying which variables are correlated, e.g., finding a link between mobility and depression; or for making inferences, e.g., using machine-learning techniques to infer physical activities from accelerometer data.
- *Ground truth*: An important aspect of sensing studies to consider is the inclusion of a reference to measure the performance of the classification or inference process. This reference, known as ground truth, could be objective, when it consists of data collected through an already validated/reliable technological instrument (e.g., a GPS), or clinical instruments (e.g., using the mini-mental state examination (MMSE)). Ground truth could also be subjective, when it is obtained through participants' self-report.

1.2.2.2 Data collection techniques

This set of characteristics was extracted to identify trends in the techniques, including the types of mobile technologies used for data collection. They help us determine what, how, and when the data collection was conducted.

- *Sensing scope*: This is related to the main focus of a sensing study (i.e., situations about the subject or about the surroundings of the subject): it can be person-centric or environment-centric. Person-centric focuses on situations about the person carrying the phone or mobile technology, by monitoring or sharing information that the person considers is sensitive [10]: for instance, location, daily life patterns, physical activities, health condition (e.g., heart rate and glucose level). In environment-centric sensing, the data collected relates to the person's surroundings, and could be shared with everyone for the public good. It involves sensing environmental data (e.g., noise, air pollution) or fine-grained traffic information (e.g., free parking slots, traffic jam information) [10].
- *Sensing paradigm*: This refers to the way the user intervenes in the study. This classification is not new, as it has been mentioned in Refs. [10,11]. There are two types of sensing paradigms: participatory sensing and opportunistic sensing. Participatory sensing means that the user actively partakes in collecting data. Users can decide when, what, and how to capture data. It can also be the case that the study suggests to users when to collect data as a call to action, so users are directly involved in the process of collection. In the opportunistic sensing paradigm, users are only the device bearers and they are not explicitly participating in data collection. The sensing system by itself makes all of the decisions according to a schedule or an event, and does not require active user participation.
- *Types of mobile sensors*: It includes mobile phone's built-in sensors (e.g. accelerometer, GPS, microphone, etc.). However, as of today no smartphone has incorporated all the sensors available in the market. In terms of space, it would be very difficult to fit them all; also, in terms of energy, it would be challenging to power them all. Indeed, there are many sensors that can communicate with smartphones via personal area networks communications protocols (e.g., Bluetooth). Therefore, we also identified external sensors custom-made to gather information about specific conditions (e.g., pollution, cycling cadence), and commercial sensors (e.g., Fitbit). Additionally, we identified if the phone was used as a hub for storing/processing/uploading data collected from external or commercial sensors.

1.2.2.3 Multidisciplinary research

- *Research areas*: Reveals the different research and domain application areas that were involved. We identified whether clinical and social sciences areas were involved in each research study, to either assess and/or conduct it. This involvement was evident through the papers' coauthoring information (i.e., researchers/practitioners from clinical areas appeared as coauthors), or if they report that an institutional review board (IRB) was formally designated to assure that the study protocol included appropriate steps to protect the rights and safety of the participants.
- *Subject area*: Additionally, each paper was classified according to the subject areas as provided by the journal [38] and conferences [39] rankings. Alternatively, for those not appearing in these rankings, we reviewed their call for papers to determine the area to which they contributed. Thus, we classified paper contributions as Computer Science/Engineering (Cs/Eng), Medical and Biological area (Med/Bio), and Social Sciences and Humanities.

1.2.2.4 Health care application domain

Finally, we also identified which papers reported studies addressing medical conditions and/or wellness issues, and classified them by using the following criteria:

- *Dimensions of wellness*: Wellness is defined as a positive outcome that is meaningful for people. It helps to know how people perceive that their lives are going well [40]. This category includes papers measuring indicators of specific living conditions (such as the quality of their relationships, stress, and environmental noise level), and/or using the collected data to make participants aware of a well-being condition. The wellness conditions were categorized in seven dimensions [41]: social wellness, physical wellness, emotional wellness, career wellness, intellectual wellness, and environmental wellness.
- *Type of disease*: Studies addressing a medical condition were classified as Communicable disease, which is caused by infectious agents such as influenza, hepatitis or HIV/AIDS; and as noncommunicable disease, which is noninfectious or nontransmissible, such as depression, obesity, cancer, or diabetes [42].

1.3 **RESULTS**

We identified 77 papers that fit the inclusion criteria: 14 of them reported a literature review or survey, and the remaining 63 were included in this review. All selected papers were published from 2003 to 2016. Eighteen papers refer to studies published in journals, 44 in conference proceedings, and 1 in a book chapter. Table 1.1 summarizes the data abstracted from the papers. Papers suitable for inclusion were counted for each category. Some categories were not mutually exclusive, such as type of mobile sensors, as some papers included more than one type (e.g., built-in sensors, external sensors). Thus, this means that some papers were considered in two or more categories: i.e., some papers were counted more than once.

1.3.1 **SENSING SCOPE AND PARADIGM**

Fig. 1.1 shows that, during the reported years, there was more opportunistic than participatory sensing. We found that 57 papers used an opportunistic approach (as opposed to 20 papers in which participatory sensing was used). However, it is important to consider that several studies, 14 to be precise, combine these two sensing paradigms, such as [21,27,29,43–53]. As seen, not many papers were published prior to 2007, when the first smartphone became available. There seems to be a valley between 2011 and 2014, which could have been the result of the search terms used. The trend in the number of papers published, however, seems to be steadily increasing. We found more studies that are person-centric (PC = 49) than environment-centric (EC = 17), with the opportunistic and person-centric combination the most used (44 papers).

1.3.2 **STUDY AIMS AND CONTRIBUTIONS**

There are 34 papers reporting the use of an objective ground truth to compare and assess their results, which is more than twice the use of subjective ground truth (14 papers). From the 20 papers categorized as not using any ground truth, 13 present a framework or toolkit, while the remaining 7 report sensing studies by using reliable commercial sensors or mobile phone built-in sensors to study the users'

Table 1.1 Characteristics of Papers Analyzed

Sensing Paradigm O: Opportunistic P: Participatory B: Both	Ground Truth O: Objective S: Subjective N: Not used			Type of Mobile Sensors B: Built-in E: External C: Commercial			Sensing Scope P: Person E: Environment		Subject Area C: Cs/Eng M: Med/Bio B: Both			Application Domain W: Wellness D: Disease O: Others			Multidisciplinary Y: Yes N: No	
	O	S	N	B	E	C	P	E	C	M	B	W	D	O	Y	N
O (43)	26	8	11	34	13	1	31	13	40	1	2	18	4	23	5	38
P (6)	3	0	3	4	2	1	5	1	6	0	0	4	2	1	2	4
B (14)	5	6	6	13	1	0	13	3	11	3	0	9	2	5	6	8
Total (63)	34	14	20	51	16	2	49	17	57	4	2	31	8	29	13	50

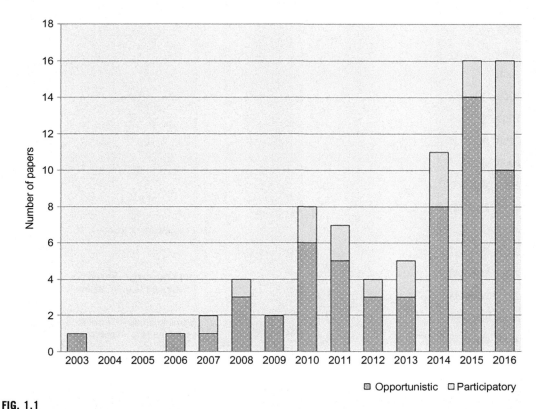

FIG. 1.1

Number of papers for opportunistic and participatory sensing paradigms.

attitudes or behavior related to a wellness condition. For instance, in Ref. [54], they use the mobile phone's GPS and body sensors to monitor the heart rate and body postures to map a cycling experience.

Also, following Fig. 1.2, it can be seen that there has been an increase of papers employing predictive models (e.g., machine learning) rather than more descriptive ones (i.e., based on statistical correlations). Perhaps not very surprising is the fact that the first papers describing a toolkit or framework for carrying out sensing studies were published in 2007, which corresponds to the year in which the IPhone was presented. The iOS, despite its popularity due to the way it handles background processes, has not been the platform of choice for developing those toolkits or frameworks. Table 1.2 presents the references to the papers reporting frameworks and/or sensing studies, categorized by the sensing paradigm that they followed.

1.3.3 HEALTH CARE APPLICATION DOMAINS

We identified 34 papers addressing a wellness and/or disease condition, while 29 pertain to other application domains, such as transportation and activity modeling. As presented in Table 1.3, 31 out of the 34 papers report sensing studies trying to infer a wellness condition, e.g., monitoring a person's stress level [25,52] or sleep [23]; and 8 papers address diseases like common cold and

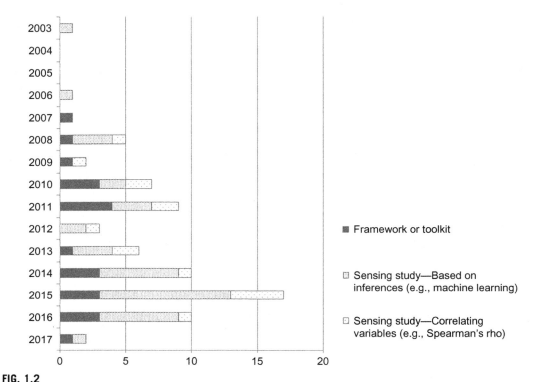

FIG. 1.2

Number of studies by *study type*.

Table 1.2 References to the 63 Papers Categorized by *Study Type* and *Sensing Paradigm* Characteristics

Study Type	Participatory	Opportunistic	Both
14 Frameworks	[55–58]	[59–65]	[43,47,48,51]
42 Sensing studies	[66,67]	[19,20,23,25,26,28,31,32,34–36,54,57,68–86]	[29,44–46,49,50,52,53]
7 Frameworks + sensing studies	[87]	[22,24,30,33]	[21,27]

influenza [50], schizophrenia [86], and dementia [57]. As seen, most of the studies on wellness focus on 4 of the 7 wellness dimensions that we used: the physical (13 papers), emotional (9 papers), environmental (6 papers), and social (5 papers).

1.3.4 MOBILE SENSORS

For the papers reported, different types of sensors were used, mainly from three sources: (a) 51 out of the 63 papers used the smartphone itself (built-in sensors), (b) 16 used custom external devices or sensors, and (c) 2 commercial external devices. Generally, the sensors that fit into these three sources are

Table 1.3 References to the 34 Papers Addressing Wellness or Disease Conditions Categorized by *Sensing Scope* and *Sensing Paradigm*

Refs.	Wellness				Disease		Sensing Scope		Sensing Paradigm	
	So	Ph	Em	Env	C	NC	PC	EC	O	P
[54]		✓					✓		✓	
[68]				✓				✓	✓	
[73]		✓					✓		✓	
[53]		✓					✓		✓	✓
[57]						✓	✓		✓	
[85]			✓			✓	✓		✓	
[86]						✓	✓		✓	
[66]				✓				✓	✓	✓
[60]		✓					✓		✓	✓
[55]		✓					✓		✓	✓
[59]				✓				✓	✓	
[46]			✓				✓		✓	✓
[76]				✓				✓	✓	
[56]						✓	✓			✓
[78]				✓				✓	✓	
[67]		✓				✓	✓			✓
[45]			✓				✓		✓	✓
[21]		✓					✓		✓	✓
[23]		✓					✓		✓	
[24]	✓	✓					✓		✓	
[25]		✓					✓		✓	
[26]			✓			✓	✓		✓	✓
[30]	✓						✓		✓	
[31]	✓						✓		✓	
[32]	✓						✓		✓	
[33]			✓				✓		✓	
[34]				✓				✓	✓	
[52]			✓				✓		✓	✓
[75]		✓				✓	✓		✓	
[79]		✓					✓		✓	
[44]		✓					✓		✓	✓
[87]			✓				✓			✓
[50]	✓		✓		✓		✓		✓	✓
[49]			✓				✓		✓	✓

So, *social*; Ph, *physical*; Em, *emotional*; Env, *environmental*; C, *communicable*; NC, *noncommunicable*; PC, *person centric*; EC, *env. centric*; O, *opportunistic*; P, *participatory*.

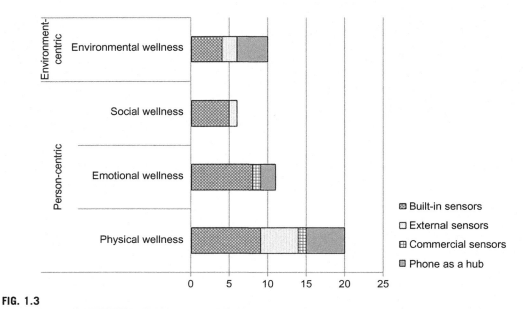

FIG. 1.3

Number of papers using different *types of sensors* in wellness-related studies by *sensing scope*.

called hard sensors, as their purpose is to measure the real world and they generally are based on the use of hardware (e.g., microchips). However, there is another type of sensor based on software: logs about the use of the mobile phone (e.g., call logs, SMS logs, internet usage logs, calendar logs, social network logs) [28,49–51,61,88]. These "sensors" virtually exist in almost every mobile phone. Usually, these sensors have been used to infer social or personality traits of the user. According to Félix et al. [6], most studies prior to 2016 used hard sensing (e.g., accelerometer, GPS, Bluetooth and microphone), since on-device hard sensors are the most accessible and less obtrusive. These statements are reinforced in this review as illustrated in the graph of Fig. 1.3, which presents how many papers, addressing a wellness condition, report using each of the types of mobile sensors. For instance, from the 6 studies on environmental wellness, 4 papers report using built-in sensors, while 2 out of 6 used external sensors, and, finally, 4 used the phone as a hub for processing data. Thus, this graph provides an appreciation of the focus of most studies. While a few of them have focused on studying environmental issues through mobile sensing, the vast majority have focused on the individual, studying aspects related to wellness such as physical, emotional, or social.

1.3.5 MULTIDISCIPLINARY RESEARCH

Even though it is evident how mobile sensing is being applied or used to study research health care issues, there are more studies published in venues or forums related to computer science and engineering [Cs/Eng = 57] than in medical and biological [Med/Bio = 4]; and only 2 papers were classified as both Cs/Eng and Med/Bio. This is associated with the fact that most of the papers do not report collaboration with health care researchers or practitioners, or include an IRB from this area.

In Fig. 1.4 it can be seen that, whereas the sensing scope was mainly person-centric, engineering-only areas tended to be more diverse in that they mainly addressed issues that go beyond the individual.

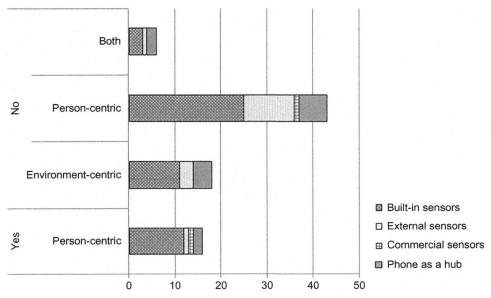

FIG. 1.4

Number of papers using different *types of sensors* in a particular *sensing scope* by studies that include clinical departments or health care researchers (yes = including clinical areas; no = only engineering areas).

As expected, the involvement of clinical areas focused mainly on the individual, i.e., person-centric sensing scope. Also, as opposed to what one might think, most studies that also include health care researchers rely on sensors that are embedded in mobile phones rather than on commercial sensors or external ones.

1.4 CONCLUSIONS

Mobile sensing is increasingly becoming a part of everyday life due to the rapid evolution of sensing platforms. Diverse applications such as social networks, health care, environmental monitoring, transportation safety, and social business can vastly benefit from the contributions of mobile sensing. This chapter discusses the current state of the art in the emerging field of mobile sensing. We have analyzed a large amount of research works and sensing applications and have summarized the observed trends in those research works.

Mobile phone built-in sensors are by far the most used types of hard sensors. This is expected, as they are now readily available in modern mobile phones. In addition, more recent papers seem to have favored the development of studies that include external sensors such as wireless electrocardiograms (ECGs) [87], electroencephalography (EEG) headsets [37] to control mobile phones, or out-of-voice-band microphones [76] used for identifying nonspeech body sounds, such as sounds of food intake, breath, or laughter. This is mainly due to new devices that allow wireless connection between the mobile phones and those devices, mainly via Bluetooth.

Most of the research on mobile sensing that studies medical and wellness conditions is being published in venues pertaining to the engineering and computer science areas; however, some multidisciplinary work is being conducted (barely). This is expected to change in the years to come due to the aforementioned precision medicine initiative, in which medical areas will be using mobile-sensing technologies to monitor patients or create much more precise characterizations of health-related conditions. Also, many of the efforts in wellness are related to either physical or emotional wellness, which makes sense, as most published papers have been person-centric studies.

REFERENCES

[1] Collins FS, Varmus H. A new initiative on precision medicine. N Engl J Med 2015;372:793–5. https://doi.org/10.1056/NEJMp1500523.

[2] Pham MT, Rajić A, Greig JD, Sargeant JM, Papadopoulos A, McEwen SA. A scoping review of scoping reviews: advancing the approach and enhancing the consistency. Res Synth Methods 2014;5:371–85. https://doi.org/10.1002/jrsm.1123.

[3] Daudt HM, van Mossel C, Scott SJ. Enhancing the scoping study methodology: a large, inter-professional team's experience with Arksey and O'Malley's framework. BMC Med Res Methodol 2013;13:48. https://doi.org/10.1186/1471-2288-13-48.

[4] Pau G, Segre Reinach S, Im M, Tolic I, Tse R, Marfia G. In: Mobile sensing and beyond in the information age. Proceedings of 2015 workshop on pervasive wireless healthcare—MobileHealth'15. Hangzhou, China: ACM; 2015. p. 3–6. https://doi.org/10.1145/2757290.2757294.

[5] Sagl G, Resch B, Blaschke T. Contextual sensing: integrating contextual information with human and technical geo-sensor information for smart cities. Sensors 2015;15:17013–35. https://doi.org/10.3390/s150717013.

[6] Félix IR, Castro LA, Rodríguez L-F, Ruiz ÉC. In: Mobile phone sensing: current trends and challenges. International conference on ubiquitous computing and ambient intelligence. Puerto Varas, Chile: Springer; 2015. p. 369–74. https://doi.org/10.1007/978-3-319-26401-1_34.

[7] Nijkamp P. Data from mobile phone operators: a tool for smarter cities. Telecommun Policy 2014;39:335–46.

[8] Contreras-Naranjo JC, Wei Q, Ozcan A. Mobile phone-based microscopy, sensing, and diagnostics. IEEE J Sel Top Quantum Electron 2016;22:392–405. https://doi.org/10.1109/JSTQE.2015.2478657.

[9] Ali S, Khusro S. Mobile phone sensing: a new application paradigm. Indian J Sci Technol 2016;9:1–42. https://doi.org/10.17485/ijst/2016/v9i19/53088.

[10] Khan WZ, Xiang Y, Aalsalem MY, Arshad Q. Mobile phone sensing systems: a survey. IEEE Commun Surv Tutorials 2013;15:402–27. https://doi.org/10.1109/SURV.2012.031412.00077.

[11] Lane N, Miluzzo E, Lu H, Peebles D, Choudhury T, Campbell A. A survey of mobile phone sensing. IEEE Commun Mag 2010;48:140–50. https://doi.org/10.1109/MCOM.2010.5560598.

[12] Calabrese F, Ferrari L, Blondel VD. Urban sensing using mobile phone network data: a survey of research. ACM Comput Surv 2014;47:1–20. https://doi.org/10.1145/2655691.

[13] Pejovic V, Lathia N, Mascolo C, Musolesi M. Mobile-based experience sampling for behaviour research. In: Emotions and personality in personalized services. Cham: Springer; 2016. p. 141–61. https://doi.org/10.1007/978-3-319-31413-6_8.

[14] Ren J, Zhang Y, Zhang K, Shen X. Exploiting mobile crowdsourcing for pervasive cloud services: challenges and solutions. IEEE Commun Mag 2015;53:98–105. https://doi.org/10.1109/MCOM.2015.7060488.

[15] Nebeker C, Lagare T, Takemoto M, Lewars B, Crist K, Bloss CS, et al. Engaging research participants to inform the ethical conduct of mobile imaging, pervasive sensing, and location tracking research. Transl Behav Med 2016;6:577–86. https://doi.org/10.1007/s13142-016-0426-4.

[16] Kantarci B, Mouftah HT. In: Sensing services in cloud-centric Internet of things: a survey, taxonomy and challenges. 2015 IEEE international conference on communications workshops. London, UK: IEEE; 2015. p. 1865–70. https://doi.org/10.1109/ICCW.2015.7247452.

[17] Guo B, Wang Z, Yu Z, Wang Y, Yen NY, Huang R, et al. Mobile crowd sensing and computing. ACM Comput Surv 2015;48:1–31. https://doi.org/10.1145/2794400.

[18] Buchanan M. Behavioural science: secret signals. Nature 2009;457:528–30. https://doi.org/10.1038/457528a.

[19] Kim TJ, Chu M, Brdiczka O, Begole J. In: Predicting shoppers' interest from social interactions using sociometric sensors. Proceedings of 27th international conference extended abstracts on human factors in computing systems (CHI EA'09). New York, NY: ACM Press; 2009. p. 4513. https://doi.org/10.1145/1520340.1520692.

[20] Choudhury T, Pentland A. In: Sensing and modeling human networks using the sociometer. Proceedings of seventh IEEE international symposium on wearable computers 2003. White Plains, NY: IEEE; 2004. p. 216–22. https://doi.org/10.1109/ISWC.2003.1241414.

[21] Consolvo S, Libby R, Smith I, Landay JA, McDonald DW, Toscos T, et al. In: Activity sensing in the wild: a field trial of ubifit garden. Proceeding twenty-sixth annual CHI conference on human factors in computing systems (CHI'08). New York, NY: ACM Press; 2008. p. 1797. https://doi.org/10.1145/1357054.1357335.

[22] Kang S, Kwon S, Yoo C, Seo S, Park K, Song J, et al. In: Sinabro: opportunistic and unobtrusive mobile electrocardiogram monitoring system. Proceedings of 15th workshop on mobile computing systems & applications (HotMobile'14). New York, NY: ACM Press; 2014. p. 1–6. https://doi.org/10.1145/2565585.2565605.

[23] Chen Z, Lin M, Chen F, Lane N, Cardone G, Wang R, et al. In: Unobtrusive sleep monitoring using smartphones. Proceedings of ICTs improving patients rehabilitation research techniques, ICST. Brussels: IEEE; 2013. p. 145–52. https://doi.org/10.4108/icst.pervasivehealth.2013.252148.

[24] Lane N, Mohammod M, Lin M, Yang X, Lu H, Ali S, et al. In: BeWell: a smartphone application to monitor, model and promote wellbeing. Proceedings of 5th international ICST conference on pervasive computing technologies for healthcare. Dublin, Ireland: IEEE; 2011. p. 8. https://doi.org/10.4108/icst.pervasivehealth.2011.246161.

[25] Lu H, Frauendorfer D, Rabbi M, Mast MS, Chittaranjan GT, Campbell AT, et al. StressSense: detecting stress in unconstrained acoustic environments using smartphones. Ubicomp 2012;351–60. https://doi.org/10.1145/2370216.2370270.

[26] Qin Y, Bhattacharya T, Kulik L, Bailey J. In: A context-aware do-not-disturb service for mobile devices. Proceedings of 13th international conference on mobile and ubiquitous multimedia (MUM'14). New York, NY: ACM Press; 2014. p. 236–9. https://doi.org/10.1145/2677972.2678003.

[27] White J, Thompson C, Turner H, Dougherty B, Schmidt DC. WreckWatch: automatic traffic accident detection and notification with smartphones. Mob Networks Appl 2011;16:285–303. https://doi.org/10.1007/s11036-011-0304-8.

[28] Chittaranjan G, Blom J, Gatica-Perez D. In: Who's who with big-five: analyzing and classifying personality traits with smartphones. 2011 15th Annual international symposium on wearable computers. San Francisco, CA: IEEE; 2011. p. 29–36. https://doi.org/10.1109/ISWC.2011.29.

[29] Grünerbl A, Oleksy P, Bahle G, Haring C, Weppner J, Lukowicz P. In: Towards smart phone based monitoring of bipolar disorder. Proceedings of second ACM workshop on mobile systems, applications, and services for healthcare (mHealthSys'12). New York, NY: ACM Press; 2012. p. 1. https://doi.org/10.1145/2396276.2396280.

[30] Rachuri KK, Mascolo C, Musolesi M, Rentfrow PJ. In: SociableSense: exploring the trade-offs of adaptive sampling and computation offloading for social sensing. Proceedings of 17th annual international conference on mobile computing and networking (MobiCom'11). New York, NY: ACM Press; 2011. p. 73. https://doi.org/10.1145/2030613.2030623.

[31] Eagle N, Sandy Pentland A. Reality mining: sensing complex social systems. Pers Ubiquit Comput 2006;10:255–68. https://doi.org/10.1007/s00779-005-0046-3.

[32] Miluzzo E, Lane ND, Fodor K, Peterson R, Lu H, Musolesi M, et al. In: Sensing meets mobile social networks: the design, implementation and evaluation of the CenceMe application. Proceedings of 6th ACM

conference embedded network sensor systems (SenSys'08). New York, NY: ACM Press; 2008. p. 337. https://doi.org/10.1145/1460412.1460445.

[33] Rachuri KK, Musolesi M, Mascolo C, Rentfrow PJ, Longworth C, Aucinas A. In: EmotionSense: a mobile phones based adaptive platform for experimental social psychology research. Proceedings of 12th ACM international conference on ubiquitous computing (Ubicomp'10). New York, NY: ACM Press; 2010. p. 281. https://doi.org/10.1145/1864349.1864393.

[34] Rana RK, Chou CT, Kanhere SS, Bulusu N, Hu W. In: Ear-phone: An end-to-end participatory urban noise mapping system. Proceedings of 9th ACM/IEEE international conference information processing in sensor networks (IPSN'10); 2010. p. 105–16. https://doi.org/10.1145/1791212.1791226.

[35] Mohan P, Padmanabhan VN, Ramjee R. In: Nericell: rich monitoring of road and traffic conditions using mobile smartphones. Proceedings of 6th ACM conference embedded network sensor systems (SenSys'08). New York, NY: ACM Press; 2008. p. 323. https://doi.org/10.1145/1460412.1460444.

[36] Zhou P, Zheng Y, Li M. In: How long to wait?: predicting bus arrival time with mobile phone based participatory sensing. Proceedings of 10th international conference on mobile systems, applications, and services (MobiSys'12). vol. 13. New York, NY: ACM Press; 2012. p. 379. https://doi.org/10.1145/2307636.2307671.

[37] Campbell A, Choudhury T, Hu S, Lu H, Mukerjee MK, Rabbi M, et al. In: NeuroPhone. Proceedings of second ACM SIGCOMM workshop on networking, systems, and applications on mobile handhelds (MobiHeld'10). New York, NY: ACM Press; 2010. p. 3. https://doi.org/10.1145/1851322.1851326.

[38] SCImago. Scimago Journal & Country Rank, http://www.scimagojr.com/journalrank.php; 2017. Accessed 15 December 2016.

[39] Clarivate Analytics. Conference proceedings citation index, http://wokinfo.com/products_tools/multidisciplinary/webofscience/cpci/; 2017. Accessed 13 December 2016.

[40] Centers for Disease Control and Prevention. Well-being concepts, https://www.cdc.gov/hrqol/wellbeing.htm; 2016. Accessed 12 January 2017.

[41] University of Wisconsin, School of Health Promotion and Human Development. 7 Dimensions of wellness self assessment 2016. http://www.who.int/management/general/self/AreYou Balancing the 7 Dimensions of Wellness.doc [accessed 10 December 2016].

[42] World Health Organization. Noncommunicable diseases, http://www.who.int/mediacentre/factsheets/fs355/en/; 2015. Accessed 10 December 2016.

[43] Aloi G, Di Felice M, Ruggeri G, Savazzi S. In: A mobile phone-sensing system for emergency management: the SENSE-ME platform. 2016 IEEE 2nd international forum on research and technologies for society and industry leveraging a better tomorrow (RTSI). Bologna, Italy: IEEE; 2016. p. 1–6.

[44] Hasan SS, Chipara O, Wu Y-H, Aksan N. In: Evaluating auditory contexts and their impacts on hearing aid outcomes with mobile phones. Proceedings of 8th international conference on pervasive computing technologies for healthcare (PervasiveHealth 2014), Oldenburg, Germany, 20–23 May; 2014. https://doi.org/10.4108/icst.pervasivehealth.2014.254952.

[45] Wahle F, Kowatsch T, Fleisch E, Rufer M, Weidt S. Mobile sensing and support for people with depression: a pilot trial in the wild. JMIR mHealth uHealth 2016;4:e111. https://doi.org/10.2196/mhealth.5960.

[46] Asselbergs J, Ruwaard J, Ejdys M, Schrader N, Sijbrandij M, Riper H. Mobile phone-based unobtrusive ecological momentary assessment of day-to-day mood: an explorative study. J Med Internet Res 2016;18:e72. https://doi.org/10.2196/jmir.5505.

[47] Perez M, Castro LA, Favela J. In: InCense: a research kit to facilitate behavioral data gathering from populations of mobile phone users. 5th international symposium on ubiquitous computing and ambient intelligence; 2011. p. 1–8.

[48] Hicks J, Ramanathan N, Kim D, Monibi M, Selsky J, Hansen M, et al. In: AndWellness: an open mobile system for activity and experience sampling. Wireless health 2010 (WH'10). New York, NY: ACM Press; 2010. p. 34. https://doi.org/10.1145/1921081.1921087.

[49] LiKamWa R, Liu Y, Lane ND, Zhong L. In: MoodScope: building a mood sensor from smartphone usage patterns. Proceeding 11th annual international conference on mobile systems, applications, and services (MobiSys'13). New York, NY: ACM Press; 2013. p. 389. https://doi.org/10.1145/2462456.2464449.

[50] Madan A, Cebrian M, Lazer D, Pentland A. In: Social sensing for epidemiological behavior change. Proceedings of 12th ACM international conference on ubiquitous computing (Ubicomp'10). New York, NY: ACM Press; 2010. p. 291. https://doi.org/10.1145/1864349.1864394.

[51] Froehlich J, Chen MY, Consolvo S, Harrison B, Landay J. In: MyExperience. Proceedings of 5th international conference on mobile systems, applications, and services (MobiSys'07). San Juan, New York, NY: ACM Press; 2007. p. 57. https://doi.org/10.1145/1247660.1247670.

[52] Bogomolov A, Lepri B, Ferron M, Pianesi F, Pentland AS. In: Daily stress recognition from mobile phone data, weather conditions and individual traits. Proceedings of ACM international conference on multimedia (MM'14). New York, NY: ACM Press; 2014. p. 477–86. https://doi.org/10.1145/2647868.2654933.

[53] Naughton F, Hopewell S, Lathia N, Schalbroeck R, Brown C, Mascolo C, et al. A context-sensing mobile phone app (Q sense) for smoking cessation: a mixed-methods study. JMIR mHealth uHealth 2016;4:e106. https://doi.org/10.2196/mhealth.5787.

[54] Oliveira DS, Afonso JA. In: Mobile sensing system for georeferenced performance monitoring in cycling. World congress engineering 2015. vol. I. 2015. p. 1–5.

[55] Afonso JA, Rodrigues FJ, Pedrosa D, Afonso JL. Mobile sensing system for cycling power output control. Cham: Springer; 2016773–83. https://doi.org/10.1007/978-3-319-43671-5_65.

[56] Comina G, Suska A, Filippini D. Autonomous chemical sensing interface for universal cell phone readout. Angew Chem Int Ed 2015;54:8708–12. https://doi.org/10.1002/anie.201503727.

[57] Castro LA, Favela J, Beltrán J, Chávez E, Perez M, Rodriguez M, et al. In: Collaborative opportunistic sensing with mobile phones. Proceedings of 2014 ACM international joint conference on pervasive and ubiquitous computing adjunct publication (UbiComp'14 Adjunct). New York, NY: ACM Press; 2014. p. 1265–72. https://doi.org/10.1145/2638728.2638814.

[58] Ren J, Zhang Y, Zhang K, Shen X. SACRM: social aware crowdsourcing with reputation management in mobile sensing. Comput Commun 2015;65:55–65. https://doi.org/10.1016/j.comcom.2015.01.022.

[59] Issarny V, Mallet V, Nguyen K, Raverdy P-G, Rebhi F, Ventura R. In: Dos and don'ts in mobile phone sensing middleware: learning from a large-scale experiment. Proceedings of 17th international middleware conference (Middleware'16). New York, NY: ACM Press; 2016. p. 1–13. https://doi.org/10.1145/2988336.2988353.

[60] Maio AF, Afonso JA. Wireless cycling posture monitoring based on smartphones and bluetooth low energy. Lect Notes Eng Comput Sci 2015;2217:1–5.

[61] Kiukkonen N, Blom J, Dousse O, Gatica-Perez D, Laurila J. In: Towards rich mobile phone datasets: lausanne data collection campaign. Proc ACM international conference on pervasive services, Berlin; 2010.

[62] Phan T, Kalasapur S, Kunjithapatham A. In: Sensor fusion of physical and social data using web SocialSense on smartphone mobile browsers. 2014 IEEE 11th consumer communications and networking conference. IEEE; 2014. p. 98–104. https://doi.org/10.1109/CCNC.2014.6866555.

[63] Liono J, Nguyen T, Jayaraman PP, Salim FD. In: UTE: a ubiquitous data exploration platform for mobile sensing experiments. 2016 17th IEEE international conference on mobile data management. vol. 1. Porto, Portugal: IEEE; 2016. p. 349–52. https://doi.org/10.1109/MDM.2016.61.

[64] Eisenman SB, Miluzzo E, Lane ND, Peterson RA, Ahn G-S, Campbell AT. BikeNet: a mobile sensing system for cyclist experience mapping. ACM Trans Sens Netw 2009;6:1–39. https://doi.org/10.1145/1653760.1653766.

[65] Afonso JA, Rodrigues FJ, Pedrosa D, Afonso JL. Automatic control of cycling effort using electric bicycles and mobile devices. World Congr Eng 2015;2015:381–6.

[66] Hu M, Che W, Zhang Q, Luo Q, Lin H. A multi-stage method for connecting participatory sensing and noise simulations. Sensors 2015;15:2265–82. https://doi.org/10.3390/s150202265.

[67] Goel M, Saba E, Stiber M, Whitmire E, Fromm J, Larson EC, et al. In: SpiroCall: measuring lung function over a phone call. Proceedings of 2016 CHI conference on human factors in computing systems. ACM; 2016. p. 5675–85. https://doi.org/10.1145/2858036.2858401.

[68] Liu X, Xiang C, Li B, Jiang A. In: Collaborative bicycle sensing for air pollution on roadway. 2015 IEEE 12th International conference ubiquitous intelligence computing, 2015 IEEE 12th International conference on autonomic and trusted computing, 2015 IEEE 15th International conference on scalable computing and communications and its associated workshops. IEEE, Beijing, China; 2015. p. 316–9. https://doi.org/10.1109/UIC-ATC-ScalCom-CBDCom-IoP.2015.67.

[69] Bovornkeeratiroj P, Nakorn KN, Rojviboonchai K. In: Boat arrival time prediction system using mobile phone sensing. 2015 12th International conference on electrical engineering/electronics, computer, telecommunications and information technology. Hua Hin, Thailand: IEEE; 2015. p. 1–6. https://doi.org/10.1109/ECTICon.2015.7207088.

[70] Vu L, Nguyen P, Nahrstedt K, Richerzhagen B. Characterizing and modeling people movement from mobile phone sensing traces. Pervasive Mob Comput 2015;17:220–35. https://doi.org/10.1016/j.pmcj.2014.12.001.

[71] Incel O. Analysis of movement, orientation and rotation-based sensing for phone placement recognition. Sensors 2015;15:25474–506. https://doi.org/10.3390/s151025474.

[72] Rachuri KK, Hossmann T, Mascolo C, Holden S. In: Beyond location check-ins: exploring physical and soft sensing to augment social check-in apps. 2015 IEEE international conference on pervasive computing and communications. St. Louis, MO: IEEE; 2015. p. 123–30. https://doi.org/10.1109/PERCOM.2015.7146518.

[73] Stisen A, Blunck H, Bhattacharya S, Prentow TS, Kjærgaard MB, Dey A, et al. In: Smart devices are different: assessing and mitigatingmobile sensing heterogeneities for activity recognition. Proceedings of 13th ACM conference on embedded networked sensors system (SenSys'15). New York, NY: ACM Press; 2015. p. 127–40. https://doi.org/10.1145/2809695.2809718.

[74] Wiese J, Saponas TS, Brush a J. Phoneprioception: enabling mobile phones to infer where they are kept. Proc CHI 2013;2013:2157–66. https://doi.org/10.1145/2470654.2481296.

[75] Rahman T, Adams AT, Zhang M, Cherry E, Zhou B, Peng H, et al. In: BodyBeat: a mobile system for sensing non-speech body sounds. Proceedings of 12th annual international conference on mobile systems, applications and services (MobiSys'14). New York, NY: ACM Press; 2014. p. 2–13. https://doi.org/10.1145/2594368.2594386.

[76] Meng C, Xiao H, Su L, Cheng Y. In: Tackling the redundancy and sparsity in crowd sensing applications. Proceedings of 14th ACM conference on embedded network sensors system CD-ROM (SenSys'16). New York, NY: ACM Press; 2016. p. 150–63. https://doi.org/10.1145/2994551.2994567.

[77] Chen D, Cho K, Han S, Jin Z, Shin KG. In: Invisible sensing of vehicle steering with smartphones. Proceedings of 13th annual international conference on mobile systems, applications, and services (MobiSys'15). New York, NY: ACM Press; 2015. p. 1–13. https://doi.org/10.1145/2742647.2742659.

[78] Rana R, Chou CT, Bulusu N, Kanhere S, Hu W. Ear-phone: a context-aware noise mapping using smart phones. Pervasive Mob Comput 2015;17:1–22. https://doi.org/10.1016/j.pmcj.2014.02.001.

[79] Kwapisz JR, Weiss GM, Moore SA. Activity recognition using cell phone accelerometers. ACM SIGKDD Explor Newsl 2011;12:74. https://doi.org/10.1145/1964897.1964918.

[80] Siripanpornchana C, Panichpapiboon S, Chaovalit P. In: Incidents detection through mobile sensing. 2016 13th International conference on electrical engineering/electronics, computer, telecommunications and information technology. Chiang Mai, Thailand: IEEE; 2016. p. 1–6. https://doi.org/10.1109/ECTICon.2016.7561238.

[81] Woodard D, Nogin G, Koch P, Racz D, Goldszmidt M, Horvitz E. Predicting travel time reliability using mobile phone GPS data. Transp Res C Emerg Technol 2017;75:30–44.

[82] Phithakkitnukoon S, Horanont T, Witayangkurn A, Siri R, Sekimoto Y, Shibasaki R. Understanding tourist behavior using large-scale mobile sensing approach: a case study of mobile phone users in Japan. Pervasive Mob Comput 2015;18:18–39. https://doi.org/10.1016/j.pmcj.2014.07.003.

[83] Sun Z, Hao P, Jeff X, Yang D. Trajectory-based vehicle energy/emissions estimation for signalized arterials using mobile sensing data. Transp Res Part D Transp Environ 2015;34:27–40.

[84] Mahmoud RA, Karameh FN, Hajj H. In: A system identification approach for recognition of personalized user motion patterns from mobile sensing data. 2016 3rd Middle East conference on biomedical engineering. Beirut, Lebanon: IEEE; 2016. p. 1–5. https://doi.org/10.1109/MECBME.2016.7745395.

[85] Saeb S, Zhang M, Karr CJ, Schueller SM, Corden ME, Kording KP, et al. Mobile phone sensor correlates of depressive symptom severity in daily-life behavior: an exploratory study. J Med Internet Res 2015;17:1–11. https://doi.org/10.2196/jmir.4273.

[86] Ben-Zeev D, Wang R, Abdullah S, Brian R, Scherer E, Mistler L, et al. Mobile behavioral sensing for outpatients and inpatients with schizophrenia. Psychiatr Serv 2015;67(5):558–61. https://doi.org/10.1176/appi.ps.201500130.

[87] Gaggioli A, Pioggia G, Tartarisco G, Baldus G, Corda D, Cipresso P, et al. A mobile data collection platform for mental health research. Pers Ubiquit Comput 2013;17:241–51. https://doi.org/10.1007/s00779-011-0465-2.

[88] Chronis I, Madan A, Sandy Pentland A. In: SocialCircuits. Proceeding of ICMI-MLMI'09 workshop on multimodal sensor-based systems and mobile phones for social computing (ICMI-MLMI'09). New York, NY: ACM Press; 2009. p. 1–4. https://doi.org/10.1145/1641389.1641390.

FURTHER READING

[1] Greenhalgh T. Effectiveness and efficiency of search methods in systematic reviews of complex evidence: audit of primary sources. BMJ 2005;331:1064–5. https://doi.org/10.1136/bmj.38636.593461.68.

DATA FUSION ARCHITECTURE OF HETEROGENEOUS SOURCES OBTAINED FROM A SMART DESK

2

Julio Muñoz-Benítez*, Guillermo Molero-Castillo*,†, Edgard Benítez-Guerrero*

University of Veracruz, Xalapa, Mexico* CONACYT, Mexico City, Mexico†

2.1 INTRODUCTION

In recent years, technology has evolved and adapted to the needs of users, to the development of new communication techniques, and to the emergence of new ubiquitous processing devices—mobile technology [1,2]. Among the emerging advances associated with technological development are perception and context management: that is, how systems can perceive and capture the world that surrounds the user in such way that the systems can adapt to provide the appropriate service.

Context is the physical, emotional, and social environment in which the user is immersed, and which gives significance, meaning, and value to the actions or activities that are carried out around him [3]. Through context, the situation of an entity, such as a person, place, or object, considered relevant for the interaction between a user and a system can be characterized [3a]. To analyze this interaction, it is important to have contextual information that answers some questions, known as the five Ws [4]: (a) who, which focuses on the user identity; (b) what, which refers to what the user is doing; (c) where, which is the user location; (d) when, associated with time; and (e) why, which includes elements of the user's emotional state.

An action of contextualizing involves putting in context a situation that is received in isolation or jointly from all those elements that surround it and that influence that action. This is known as context-aware computing, which detects the activity of the user and reacts to changes, providing services that are useful in the performance of the user's daily activities. To carry out this detection and reaction, different data sources must be analyzed [1]. These sources can be sensors that measure a physical variable, such as accelerometers for measuring displacement and position, photodiodes for measuring proximity and light intensity, and thermoresistances for measuring temperature, and logical sources that provide nonphysical information, such as user agenda, preferences, and configurations, among others. In general, these data sources are heterogeneous due to the variety of formats and sample rates, so these characteristics make the data obtained completely different [2].

A complex activity, prior to the analysis of the detection and reaction in a given context, is data fusion from heterogeneous sources [5]. This data fusion is the process of detecting, associating, correlating, estimating, and combining data at various levels [6]; the data come from different sources, such as sensors and databases [7], signs and decisions [8], and even human observations [9] and

Intelligent Data Sensing and Processing for Health and Well-being Applications. https://doi.org/10.1016/B978-0-12-812130-6.00002-0

experience [10]. This knowledge field has been used in diverse areas, such as [6,7,9,11]: signal processing, information theory, estimation, statistics, inference, and artificial intelligence, making greater advances in military applications through recognition of automatic targets, autonomous vehicle navigation, remote sensing, and threat identification. Other nonmilitary applications are industrial process monitoring [12], robotics [13], and medical applications [14], among others.

Data fusion has also been used to integrate various sources for detecting and classifying activities in intelligent houses [15], immersive virtual environments [16], tangible interfaces [17], and smart desks [18]. Precisely, in tangible interfaces and smart desks, due to the direct interaction that the user has with the digital system through the manipulation of objects, a wide variety of heterogeneous data is produced due to the use of multimodal sensors, objects, and applications whose main purpose is to facilitate the efficiency of user work, adapting to changes in context or situation characterization [9,19]. Data fusion of these heterogeneous sources is a major challenge because of the wide range of sensors that are used to perform dynamic interaction of the user, objects, and environment [1,2].

Undoubtedly, the interest in data fusion is increasing due to the incorporation of sensors in computing systems and devices; the objective is to have useful information as support in the decision-making process over a particular event, object, or action. This chapter presents the fundamentals of data fusion and its application in contextual information analysis. An analysis is carried out of the current models, architectures, and methods used in data fusion. In addition, based on the background of current literature, a conceptual projection of a data fusion architecture for heterogeneous sources, through data fusion methods according to the types of sources obtained, is proposed.

2.2 LITERATURE REVIEW

One of the first works related to data fusion methods dates back to 1786 with the method of the Marquis of Condorcet, mathematician and philosopher of the 18th century, who used this approach during elections [20]. Subsequently, data fusion methods were applied in various disciplines, such as reliability [21], pattern recognition [22], artificial neural networks [23], decision-making processes [8,24], statistical estimation [25,26], and weather prediction [27]. In addition, one of the areas that has given greater impetus to data fusion is data science, where methods are used to extract, transform, and load data sources as part of the data engineering process, prior to the analysis of these [28,29].

In systems with various sensors, data fusion methods have proven to be especially useful because they provide the system with the ability to use data from multiple sources. Data fusion is a key and critical issue in systems with diverse sources (sensors, users, and others). The goal is to fuse or combine data efficiently and accurately from multiple sources to overcome the limitations of using a single source [2].

In the specific case of context-aware computing, where data is obtained from various sources that attempt to describe the actions that are generated by the system and user, there is an extensive network of sensors distributed logically, spatially, or geographically in an environment and connected by a transmission network. Sensors can be visual (cameras), auditory (microphones), infrared, and sensors to measure humidity and temperature, among others. There are advantages derived from using systems from multiple sources versus traditional systems [30,31].

An example of data fusion in a context-aware system is presented in Fig. 2.1, where data is obtained from various sensors (S_1, S_2, ..., S_n), as well as from various sources of information, such as databases, data warehouses, blogs, and calendars, among others. The acquired data is fused and through queries is

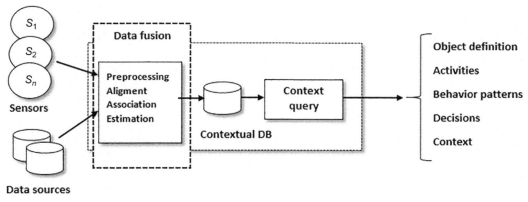

FIG. 2.1

Data fusion in a context-aware system.

Based on Benítez-Guerrero E. Context-aware mobile information systems: data management issues and opportunities.
In: Proceedings of the international conference on information & knowledge engineering, Las Vegas, NV, July 12–15; 2010.

Table 2.1 Classification Based on Input and Output Data

Classification	Input	Output
DAI-DAO	Data	Refined data
DAI-FEO	Data	Features
FEI-FEO	Features	Refined features
FEI-DEO	Features	Decisions
DEI-DEO	Decisions	Refined decisions

compared with a contextual database, to obtain a certain context inference. This inference provides information about the identity of the object, user, and behavior patterns, or even allows the system to make decisions adapting itself to a certain situation.

Dasarathy [8] proposes a classification based on input and output data (Table 2.1). This classification allows fusion from different sources to facilitate the inference of the context at a given moment, that is, the input information (data) is integrated to obtain refined data or features that describe the actions of the user and his behavior, or the environment's situation; then these features, as input data, are fused to obtain refined features or decisions; and these decisions translate into actions for better decision-making.

There are several models for data fusion implementation, such as: (a) Joint Directors Laboratories (JDL) [7]; (b) Thomopoulos [32]; (c) Multisensor integration [32a]; (d) Based on behavioral knowledge [33]; (e) Waterfall [34]; and (f) Omnibus [35]; and architectures, such as centralized, decentralized, and distributed [7], which define the levels of complexity, processes, and the moment when the data must be fused.

One of the most important decisions when designing data fusion systems is precisely where the fusion process will take place. The selection of adequate architecture varies depending on the

Table 2.2 Advantages and Limitations of Data Fusion Architectures

Architecture	Description	Advantages	Limitations
Centralized	The fusion node resides in the central processor that receives information from all sources	As it is only a central node, the fusion process obtains only the necessary data	Greater bandwidth is required to send data to the central process, which can cause bottlenecks, creating delays
Decentralized	Each node has fusion capabilities, that is, each node performs the fusion of data obtained by itself	Data fusion is done autonomously in each of the nodes	It demands a high computational cost. Communication problems arise as nodes increase
Distributed	Data is processed at each node independently before the information is sent to the fusion node	Each node provides data that is aligned and estimated before being fused; this fused data together with data from other nodes provides a global view of the situation	Loss of information may occur between fusion nodes

requirements, organization, computing resources, type of data, existing node network, communication bandwidth, desired precision, and the capacity of the sensors [7,20,36]. Based on the reviewed literature, three types of architecture that support the fusion process were identified (centralized, decentralized, and distributed). Table 2.2 presents a summary of the advantages and limitations of the mentioned architectures.

However, due to the nature of the data, the wide variety of sensors and the lack of an ideal data fusion algorithm, these models and architectures are not enough, so Khaleghi et al. [37] propose a classification of current challenges identified in data fusion (Fig. 2.2). These challenges are categorized according to the nature of data, such as (a) imperfection; (b) correlation; (c) inconsistency; and (d) disparity; each one includes subcategories with more specific problems.

- *Imperfection*: Data supplied by the sensors can be affected by a certain level of imprecision, and in the same way by the uncertainty in the measurements. This uncertainty is presented not only by the imprecision or noise of sensor measurements but also by the ambiguities and inability of the fusion system to distinguish them.
- *Correlation*: This type of problem is common in distributed configurations, such as wireless sensor networks, where some of the sensors obtain the same data from different routes or due to cyclic paths of data flow.
- *Inconsistency*: This is one of the most important problems in data fusion due to the inherent uncertainty of the sensor measurements, obtaining incorrect data. The reasons for obtaining these incorrect data may be related to permanent failures or short-term failures in measurements.
- *Disparity*: Data must be transformed locally by each sensor in a common format, before the fusion process. This disparity problem often occurs in sensors due to calibration errors.

These challenges arise from the varied format, type, and speed of sampling of the sensors, or the diversity and imperfection of the data to be fused [37]. Data coming from sensors are affected by a certain degree of imprecision and uncertainty in the measurements. This generates inconsistency in the data. In addition, the sources may have different communication speeds, which causes the data to be sent in

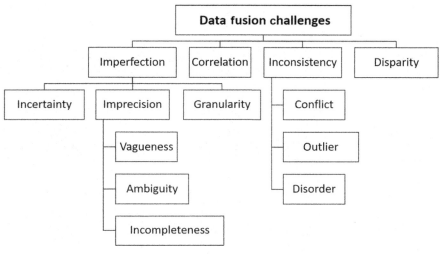

FIG. 2.2

Taxonomy of data fusion challenges.

Modified from Khaleghi B, Khamis A, Karray F, Razavi, S. Multisensor data fusion: a review of the state-of-the-art. Inf Fusion 2013;14(1):28–44.

various ways without keeping the order at the moment of being received and which may cause outliers due to unexpected situations, measurement failures, and erroneous measurements. These data must be isolated or treated to avoid errors during the fusion process.

Works that deal with atypical data focus on the identification of these data in order to eliminate it before the fusion process; part of this is achieved with the validation of sensors [38]. Research that deals with atypical data focuses on the identification or prediction of these to eliminate them before the fusion process. This is partly achieved with the validation of sensors [38]. However, the problem with these techniques is the dependence on the prior knowledge of this information. Dempster-Shafer evidence theory [39] and Bayesian probability [7,39a] are used to solve the conflict in data [37]. In the case of data disorder, these are stored until all the data is received for a reprocessing [40], which demands a high level of management and processing.

In this sense, data fusion aims to obtain a better-quality result, from multiple sensors and varied sources, generally heterogeneous, making combinations that may not be possible from a single source. Currently, data fusion has multiple applications in the military field, in reconstitution of images, in medical diagnosis and ultimately in intelligent environments, especially in tangible interfaces and smart desks.

2.3 DATA FUSION CLASSIFICATION

Data fusion can be classified according to the following criteria [20,30,41]: (a) based on the relationships between input data; (b) based on input and output data; (c) based on levels of data abstraction, that is, measures, signals, characteristics, or decisions; and (d) based on the different types of architectures. Another data fusion classification is according to the following categories [8,20,41]:

- Low-level fusion, where raw data is used as input for more accurate data (noise reduction).
- Medium-level fusion, where the characteristics of shape, texture, or position are fused to obtain characteristics to be used in other tasks.
- High-level fusion, where decisions are fused to obtain decisions or greater precision decisions.
- Multiple-fusion, where data from different levels of abstraction are combined, for example, a fused signal with a characteristic to provide a decision.

Thus, the use of heterogeneous data sources provides significant advantages, but these sources may have different types of resolution, accuracy, ranges, and formats. Therefore, data fusion methods are required to combine the various sources and obtain an extended description of an event or object, providing a formalization of data integration. For this, you need a data fusion system that considers the following elements:

- A fusion architecture appropriate to the case of study.
- A data processing to extract the maximum amount of information.
- A working environment that does not affect the fusion process.
- Appropriate algorithms or techniques for their implementation.

On the other hand, associated with the data fusion classification there are several models for the implementation of data fusion systems for new developments. These models are used to select the algorithm or method according to the situation or need to fuse or integrate the data [36]. Among the best-known models are

- JDL (Joint Director of Laboratories) model [7]. This model is one of the most used. It identifies the process, functions, and techniques applicable to data fusion. The fusion levels are: (a) assigning the appropriate process to the type of data collected, keeping information useful for the following levels; (b) using data from the previous levels, combining spatial, parametric, and entity information to obtain a refined representation of the object; (c) evaluating information according to the observed events and data obtained; (d) making inferences about the identified object, and assessing the impact of the activities or objects detected at the previous level; and (e) a metaprocess performing four functions: monitor the process in real time, identify necessary information, determine if there is any specific requirement, and allocate data sources to achieve specific goals.
- Thomopoulos model [32]. This model is composed of three modules, which fuse the data at different levels: (a) signal fusion, which is performed at the level of the parameters obtained for the phenomenon observed; (b) evidence fusion, in which data are combined at different levels of inference based on statistical models and user-given assessment; and (c) dynamic fusion, which is done by using a mathematical model that combines the data depending on the application. These data fusion levels can be applied sequentially or exchanged according to the application, as required. Factors such as delay in transmission and errors in channels of communication, as well as the spatial and temporal alignment of the data, must be considered in the fusion system.
- Multisensor integration model [30]. This model is based on the integration of data from several sensors obtained from various sources that are combined within the hierarchically embedded fusion models. A distinction between integration and data fusion is that the former refers to the use of information from various sensors of a certain task, while the latter is any state during the integration process where the data are combined. Results are sequential in a hierarchical way.

- Model based on behavioral knowledge [33]. This model comprises a number of basic states that must be completed before the established output. Features are extracted from the data obtained from the sensors, which are aligned and associated with the other features. Fusion of these data is done at the level of the sensor.
- Waterfall model [34]. This model operates from the data level to the decision level. The system is constantly updated based on the feedback sent from the decision module. This feedback is used to calibrate or configure the system as required. In this model, there are three levels of representation: (a) acquired data is processed to obtain the necessary information; (b) features are extracted and fused to obtain a certain level of inference about the data; and (c) objects or events are obtained according to the information.
- Omnibus model [35]. Omnibus is a hybrid model that contemplates four main modules: (a) signal processing, (b) pattern recognition, (c) decision-making, and (d) control resources. These modules are used to perform tasks in data fusion, combining four elements: observe, guide, decide, and act, coupled with the model proposed in Dasarathy [8] that provides three basic levels of data fusion: data, features, and decisions.

The JDL model should be seen as a base for designing an ad hoc model [7]. In Bedworth and O'Brien [35] a combination of two or more models is proposed, to obtain a new one that fulfills the function of combining the input data, according to the type of sensors and the output to a higher level of abstraction. However, the process of choosing a model or a new design must consider the difficulties in applying data fusion. These difficulties are:

- Sensor diversity (synchronization, location, and others).
- Data diversity (image, text, spatial, physical, and other magnitudes).
- Sensor calibration when errors occur.
- Sensors limitations.
- Configuration of the sensor network.
- System feedback.

2.4 **DATA FUSION ARCHITECTURES AND METHODS**

As mentioned earlier, one of the important decisions when designing a data fusion system is where the fusion should take place. The selection of the appropriate architecture varies depending on the requirements, organization, processing capacities, data types, and networks of existing nodes. Based on this questioning, in the current literature, three types of architecture have been identified that support the data fusion process: (a) centralized, (b) decentralized, and (c) distributed.

- Centralized architecture. In this type of architecture, the node that performs the fusion resides in a central processor. The sources (nodes) obtain the measurements and transmit them to the central processor where the fusion is performed. Fig. 2.3 shows how each node obtains information that is transmitted to the fusion node, which is in charge of the alignment, association, and estimation. One of the disadvantages of this architecture is the wide bandwidth required to send data to the central node. This becomes a big problem in applications that require sending large amounts of information, creating bottlenecks and increasing response times in sending and receiving data.

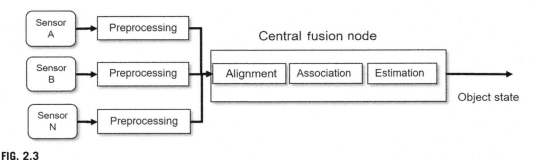

FIG. 2.3

Centralized architecture.

Based on Hall D, Llinas J. An introduction to multisensory data fusion. Proc IEEE 1997;85(1):6–23.

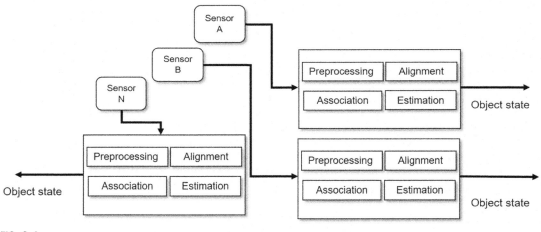

FIG. 2.4

Decentralized architecture.

Based on Hall D, Llinas J. An introduction to multisensory data fusion. Proc IEEE 1997;85(1):6–23.

- Decentralized architecture. This architecture is made up of a network of nodes with their own processing capabilities. Each node is responsible for fusing the data. There isn't a central node where fusion is performed. The process is performed autonomously: that is, the nodes communicate information obtained on the state of the object (Fig. 2.4). The main disadvantage is the high cost of communication, which increases with each node that is incorporated into the network, having the problem of scalability.
- Distributed architecture. In this architecture, the measurements of each node are processed independently before being sent to the node performing the fusion. Association of the data is carried out at each node, so the object estimation is based only on the measurements that are obtained locally; this information is sent to the node in charge of the fusion (Fig. 2.5). This architecture provides options ranging from the use of a single fusion node to the integration of intermediate fusion nodes.

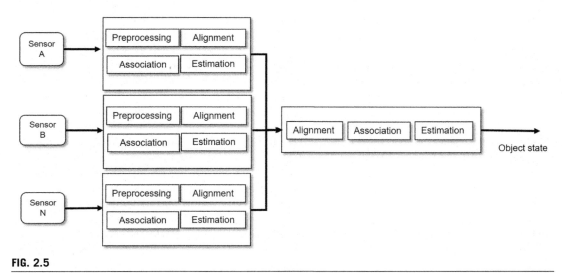

FIG. 2.5

Distributed architecture.

Based on Hall D, Llinas J. An introduction to multisensory data fusion. Proc IEEE 1997;85(1):6–23.

Centralized architecture is theoretically optimal for data fusion, assuming that alignment and association are correct and that the transmission time is not an important factor. These situations usually don't occur in real situations. One of the main problems of this architecture is the wide bandwidth required for data transmission; this problem restricts its use in most of the current applications. In the case of decentralized architecture, its main advantage is performing the data fusion in each node. This makes it ideal for covering large areas or in applications that require an extensive topology. However, its main disadvantage is the high computing cost while increasing the number of nodes; this prevents scalability of the architecture. On the other hand, a distributed architecture presents significant advantages, such as being able to distribute the processing load of data. This allows modular designs: that is, it provides different possibilities for locating one fusion node or several intermediate nodes. This makes the architecture ideal for any type of application.

In this sense, the goal of the data fusion architecture is to perform the data fusion process. That is, if two or more sources are similar, the data fusion process can be simple, but if the sensors are different, the collected data must be transformed into a common format and aligned to be processed. However, data imperfection is one of the fundamental problems of data fusion systems [37]. So, data fusion methods seek to resolve the uncertainty, vagueness, and ambiguity of the data. Two types of data fusion methods are identified [42]: (a) probabilistic, such as nearest neighbors, Bayesian approximation, state space models, conditional probability, Kalman filter, and evidential theory; and (b) statistical, such as weighted combination, multivariate statistical analysis, and arithmetic mean. Other data fusion methods are [43]: algorithms of artificial intelligence, fuzzy logic, and genetic algorithms.

- *Nearest neighbors*: This consists of clustering the data that are closest to each other, but in a dense environment it can lead to many pairs with the same probability and generate errors [43a]. If false measures occur frequently, that is, there is noise in the data, the performance of this method is diminished. A variant of this method is k-means, which consists of clustering the data set into k distinct clusters, looking for the best location of the centers of each cluster. K-means has drawbacks: it does not guarantee that the best solution to the location of the centers is found, and it is necessary to know the number of clusters.

- *Bayesian inference*: This method allows the fusion of different data, calculating a later state and a current state that is being observed; this can be updated whenever there is new data derived from a new observation. However, it presents some problems [7]: difficulty in defining the value of probabilities, complexity when the events depend on conditions, and inability to describe uncertainty about decisions.
- *Kalman filter*: This is one of the most popular methods of data fusion due to ease of implementation and a high degree of error detection. This method uses previous state estimates to make a prediction of the following states. However, this filter is sensitive to erroneous data due to the reliance on this data to obtain future estimates [20]. There are variations of this fusion method, such as the extended Kalman filter (EKF) and the unscented Kalman filter (UKF). These extensions are used in nonlinear systems [37].
- *Monte Carlo simulation*: This is a statistical model used to approximate complex and costly mathematical expressions [44]. Techniques based on this simulation, such as Monte Carlo sequence or Monte Carlo Markov chains, are the most popular approaches to probability approximation [20]. These techniques are flexible because they do not make assumptions about the probability approximation. They are based on the use of a sampling with data of greater weight and an approximation of density probability [37].
- *Dempster-Shafer theory*: This focuses on the credibility that is assigned to an event that may or not have occurred. It assumes the existence of probability values associated with determined events independently of the actual probability value [39]. This theory allows each data source to fusion at different levels. In this way, no probabilities are assigned to unknown propositions [20].
- *Fuzzy logic*: Consists of a method of data association that considers all possibilities without making an exclusive association of decisions. It is applied for taking a value within a set of values that oscillates between two extremes: absolute truth and total falsehood. It allows treating imprecise information in terms of fuzzy sets that combine rules to define actions [37,45].

Current methods must express the data in a structure suitable for analysis and processing because the data collected usually do not have the same format. One way of organizing the methods according to the data fusion problem to be solved is shown in Fig. 2.6 (data association, state estimation, and decision fusion).

Given the interest of analyzing data from heterogeneous sources, a data fusion model is needed that considers the following factors [7]:

- Appropriate methods for its implementation.
- A fusion architecture appropriate to the case of study.
- Data processing to extract the maximum amount of information.
- Precision in the fusion process.
- Optimization of the fusion process.
- Prevention of the environment from affecting the fusion process.

Despite the variety of existing methods, they still have shortcomings that can be solved with hybrid designs [35]. Therefore, one of the current challenges in merging data from different sources is to

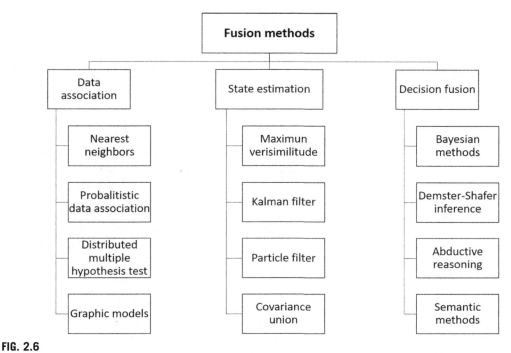

FIG. 2.6

Classification of methods according to the data fusion problem to be solved.

define appropriate methods without discarding valuable information, especially from sources with temporal and spatial correlation [46].

2.5 PROPOSAL

As mentioned in previous sections, a key aspect in data fusion is the ability of fusion methods to maintain data consistency, avoiding conflicts, outliers, and disorder. Based on the taxonomy of Khaleghi et al. [37] on the current challenges in data fusion, it should be noted that given the diversity of data sources, the fusion process cannot correct errors before data preprocessing: that is, prior to the fusion process one must contemplate resolving or reducing inconsistency errors in the data (conflicts, outliers, and disorder). Therefore, covering this particular challenge in data fusion is not trivial, since a wide variety of heterogeneous data is currently produced due to the use of multimodal sensors, objects, and applications that aim to facilitate the work efficiency of the user, adapting the system to the changes of context and characterization of a given situation. Thus, fusing data from multiple sources is a critical aspect of a system with various sensors, users, and other actors.

Current works have certain limitations, such as the manual detection of inconsistency [47] or basing this detection on the quality of the sources instead of the data [48], not to mention that inconsistency detection is carried out with different sources that share a certain nature (type, format, volume). Other works estimate the probability of receiving erroneous data based on the sensor certainty [49] or sources [48,50]. However, these works are focused on detecting and eliminating inconsistent data prior to the fusion process: that is, they lack a method for dealing with inconsistency throughout the whole process (extraction, preprocessing, fusion, and loading). Because of this, data processing must take place, prior to the fusion process. This takes place in order to reduce inconsistency and detect erroneous values. Subsequently, the consistency of the data must be maintained during the fusion process, using the appropriate algorithms or methods.

In the specific case of the smart desk, in the real world the physical manipulation plays an important role in understanding and affecting the environment. However, conventional graphical user interfaces used in distributed interaction systems do not provide a means for this type of interaction. The aim of intelligent interfaces and smart desks is to increase the physical sensation when the user interacts with the system, in order to increase collaborative work with other users, optimize the work that the users are doing, and represent interactions that the system may have to the user.

This represents a new approach to human-computer interaction, which emphasizes the physical manipulation of objects that are detected by the use of various sensors: cameras, decoders, proximity sensors, accelerometers, transceivers, among others. The system reacts to these interactions that represent digital information and adapts its behavior as necessary, even creating physical interactions that respond to those initiated by the user. In this sense, the use of sensors and users are of vital importance for these interactions to be carried out adequately, so one must have a data fusion scheme to exploit these benefits.

This work proposes, as an initial approximation, the conceptual design of a data fusion architecture of varied sources by means of integration methods according to the nature of the data, aiming to reduce the inconsistency of heterogeneous data. This analysis may be used to obtain higher quality data or even contextual inferences. Some identified data sources of the user-smart desk interaction are: (a) actions performed by the user in a given workspace; (b) the environment in which the user is located; and (c) physiological measurements of the user (breathing rate, heart rate, temperature, among others). Fig. 2.7 shows a conceptual design of an architecture of data fusion according to the nature of the data: that is, sensors, databases, logs, configurations, preferences, interactions, to mention some of these. This distributed architecture has as its main characteristic the maintenance of the consistency of the data with the lowest computational cost, through four stages of processing: (a) extraction, (b) preprocessing, (c) fusion, and (d) load.

In the extraction stage, the data are obtained from various sources according to their nature, such as sensors, databases, logs, actions, physiological measurements, among others. Some types considered are: (a) structured, which may have been stored in databases or tables (matrices, arrays, logs, and others); (b) unstructured, which is a data type with no specific format, such as text documents, raw sensor data, social networks, and others; (c) semistructured, or data that are not limited to certain fields or tables, but which have a way of separating the elements that compose them, such as XML (eXtensible Markup Language), JSON (JavaScript Object Notation), and others; and d) linear data, which may be composed of plain text, such as TXT (delimited text file), CSV (comma-separated values), and others.

The preprocessing stage is a key stage prior to the data fusion phase, since the data are prepared to be fused and some of the inconsistencies can be solved or minimized at this stage. For this, four general tasks have been defined: (a) cleaning, which consists of cleaning or debugging to avoid incomplete, duplicate or conflicting data, in order to obtain higher quality data and minimize inconsistency of data; (b) integration, which consists of clustering the data coming from different sources, although these

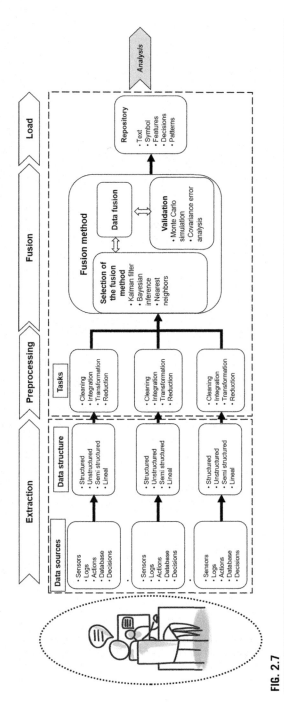

FIG. 2.7

Conceptual design of a data fusion architecture.

have different formats, avoiding problems of representation and codification; (c) transformation, which consists of transforming numerical values to nominal ones or vice versa, through normalizations, discretizations or derivations; and (d) reduction, which consists of reducing the high dimensionality of the data, which may be due to redundant data.

The fusion stage is based on the selection of one or more suitable algorithms. The selection of the algorithm depends on the nature and type of the data, such as: data attributes, symbols, patterns, or decisions. These fusion algorithms can be used at different levels of abstraction, allowing the retrieval of refined data. Some algorithms chosen for its implementation are: (a) Bayesian inference, for the representation of a current state and a later state, whenever new data are derived from a new observation; (b) nearest neighbors, to associate close data with each other, without requiring a high level of processing; and (c) Kalman filter, to make estimates of previous states, and make estimates of following states. In addition, to validate data fusion, it is proposed to implement the following methods: (a) Monte Carlo simulation, to analyze possible results by replacing values with uncertainty; and (b) analysis of covariance error, to know the effect of a categorical independent variable on a quantitative dependent variable.

The load stage is responsible for managing the data storage in a repository, warehouse, files, or other means according to the needs of its use (text, symbol, characteristics, decisions, or patterns); data analysis to detect patterns of user behavior, roles, profiles, emotions, configurations, collaborative work, and other topics of interest associated with context-aware systems; and the domain, which has proven to be useful for perceiving, capturing, and analyzing data from diverse domains, such as commercial, biomedical, environmental, and military, among others.

Since one of the major advantages of distributed architecture is the modularity, it may be implemented in a wide range of applications that require intermediate fusion modules, such as extraction, preprocessing, fusion, and load. The purpose is to homogenize heterogeneous data, making transformations and combinations by levels of fusion until the reduction or elimination of inconsistency: that is, avoiding conflicts, isolating insignificant data, and reducing disorder. Therefore, the implementation of a distributed architecture will balance the workload during the fusion process; this allows adequate control of the techniques used in data integration as required by the abstraction level of the input data.

Based on the reviewed literature, strategies whose objective is to avoid inconsistency in the data fusion process were identified. Table 2.3 summarizes strategies that aim to identify and eliminate

Table 2.3 Strategies Considered for the Implementation of the Data Fusion Architecture

Inconsistency	Problem	Characteristics	Solution Strategy
Outlier data	This data needs to be analyzed before isolation. Incorrect data fused with correct data may lead to inaccurate estimates	Identification and removal of incorrect data	Use sensor validation techniques and adaptive modeling
Disorder	Propagation time may vary, creating different sequences while sending data	Assume delays and dynamic data during analysis	Ignore, reprocess, or use prediction to incorporate delayed data
Conflict	The varied nature of the data sources, ambient noise, and sensor faults may corrupt data	The solution must be based on the type of input data	Apply corrections to data based on their nature. This means having a correction for each type of data

inconsistent data prior to the fusion process. For example, as a solution strategy in the case of outlier data, which can be caused by noise in the signal or peaks during the measurement of a sensor, the use of sensor validation techniques using adaptive models is proposed: that is, assigning weights of reliability to the sensors in order to detect reliable sources and erroneous sources. In the case of data disorder, which occurs due to the varied speeds of propagation times in sensors and heterogeneous sources, the solution strategy is to ignore, reprocess, and use predictions to incorporate ordered data into the fusion process, based on past observations or similar data states. Similarly, the nature of the data and the varied sources can create conflicts in the data that must be addressed before the fusion process, so that correction and transformation techniques should be used to normalize the data in order to detect those inconsistencies which should be avoided.

As mentioned before, the strategies used need prior information about the data, which may not be available in real time; in addition, sensor settings make the selection of validation techniques difficult. Thus, in the implementation of the proposed architecture, the objective is to develop probabilistic methods of data fusion that can overcome these limitations.

Therefore, this proposal of a distributed architecture is a first stage of the research work on the fusion of data from heterogeneous sources obtained from user activity and the use of smart desks, such as the actions performed by the user in a workspace, the environment in which the user is located, and physiological measurements (respiratory and heart rate, temperature, among others). As a future work, the implementation of this architecture is planned, divided into four stages of processing; in the fusion stage, we will develop probabilistic methods, such as Bayesian inference, Kalman filter, and nearest neighbors, whose main characteristics are the handling of conditional probability to detect the inconsistency of data not only in the sources but also in the data itself.

2.6 **CONCLUSIONS**

Data fusion has become increasingly important due to the incorporation of sensors in computer systems and the need to make these systems interact with the user and the environment, making them adaptable to situations to give adequate service when required.

The main advantage of fusing data from diverse sources, compared to having a single source, is to have more information to characterize a situation or activity that is observed. However, to have different sources it is necessary to use different sensors, which can vary in nature, format, and speed, and can even interact with logical sources, such as databases, logs, or user preferences.

The heterogeneity of data brings challenges that must be solved; the imperfection in the data, the dynamism of the environment, and the very nature of the data create inconsistencies that must be solved by suitable methods.

It is necessary to emphasize that solving the challenge of the inconsistency of data in the process of fusion is not trivial, due to the wide variety of heterogeneous data that are produced through sensors, objects, and applications. These are useful for analyzing the context and the characterization of a given situation.

Derived from the first stage of the research work on the fusion of data from heterogeneous sources, this chapter proposes the conceptual design of a data fusion architecture, structured in four stages: (a) extraction, (b) preprocessing, (c) fusion, and (d) load. The approach of this proposal is to detect inconsistencies in the data prior to the fusion process, to maintain its consistency during integration, avoiding conflict, disorder, and errors in the data.

Future work includes the implementation of the data fusion architecture, which includes fusion algorithms and data fusion evaluation methods. This will be done in order to maintain the consistency of data throughout the fusion process. For this, heterogeneous data obtained from user activity and a smart desk will be used.

Finally, defining and designing new data fusion architectures are currently major challenges in improving the accuracy of data fusion systems, which can be generated by user interactions with the system and the environment. This represents a field of opportunity for research on context-aware systems and human-computer interaction, with the purposes of improving the collaborative work among users, optimizing the tasks performed, supporting the decision-making process, and making inferences, among others.

REFERENCES

[1] Gellersen H, Schimidt A, Beigl M. Multi-sensor context-awareness in mobile devices and smart artifacts. Mob Netw Appl 2002;7(5):341–51.
[2] Olson J, Kellogg W. Ways of knowing in HCI. New York: Springer Science & Business; 2014. p. 484. ISBN: 1493903772.
[3] Benítez-Guerrero E. Context-aware mobile information systems: data management issues and opportunities. Proceedings of the international conference on information & knowledge engineering, Las Vegas, NV, July 12-15; 2010.
[3a] Wu H, Siegel M, Ablay S. Sensor fusion for context understanding. Proc. 19th instrumentation and measurement technology conference, Vol. 1. 2002;13–7.
[4] Abowd G, Mynatt ED. Charting past, present, and future research in ubiquitous computing. ACM Trans Comput Hum Interact 2000;7(1):29–58.
[5] Bernardos A, Tarrio P, Casar J. In: A data fusion framework for context-aware mobile services. IEEE international conference on multisensor fusion and integration for intelligent systems, Seoul, 20-22 August; 2008. p. 606–13.
[6] White F. Data fusion lexicon (Inf. Tec.). DTIC document, Available from www.dtic.mil/cgi-bin/GetTRDoc?Location=U2&doc=GetTRDoc.pdf&AD=ADA529661; 1991. Accessed September 2016.
[7] Hall D, Llinas J. An introduction to multisensory data fusion. Proc IEEE 1997;85(1):6–23.
[8] Dasarathy B. Sensor fusion potential exploitation-innovative architectures and illustrative applications. Proc IEEE 1997;85(1):24–38.
[9] Jenkins M, Gross G, Bisantz A, Nagi R. Towards context aware data fusion: modeling and integration of situationally qualified human observations to manage uncertainty in a hard+ soft fusion process. Inf Fusion 2015;21:130–44.
[10] Bogaert P, Gengler S. Bayesian maximum entropy and data fusion for processing qualitative data: theory and application for crowdsourced cropland occurrences in Ethiopia. Stoch Env Res Risk A 2017;31:1–17.
[11] Lee P, You P, Huang Y, Hsieh Y. Inconsistency detection and data fusion in USAR task. Eng Comput 2017;34(1):18–32.
[12] Kabir G, Demissie G, Sadiq R, Tesfamariam S. Integrating failure prediction models for water mains: Bayesian belief network based data fusion. Knowl-Based Syst 2015;85:159–69.
[13] Simanek J, Reinstein M, Kubelka V. Evaluation of the EKF-based estimation architectures for data fusion in mobile robots. IEEE Trans Mechatron 2015;20(2):985–90.

[14] Koshmak G, Loutfi A, Linden M. Challenges and issues in multisensor fusion approach for fall detection. J Sens 2015;2016:6931789, https://doi.org/10.1155/2016/6931789.

[15] Rodriguez S, De Paz J, Villarrubia G, Zato C, Bajo J, Corchado J. Multi-agent information fusión system to manage data from a wsn in residential home. Inf Fusion 2015;23:43–57.

[16] Surie D, Pederson T, Lagriffoul F, Janlert L, Sjölie D. In: Activity recognition using an egocentric perspective of everyday objects. International conference on ubiquitous intelligence and computing, Hong Kong; 2007. p. 246–57.

[17] Branton C, Ullmer N, Wiggins A, Rogge L, Setty N, Beck SD, et al. In: Toward rapid and iterative development of tangible, collaborative, distributed user interfaces. Proceedings of the 5th ACM SIGCHI symposium on engineering interactive computing systems, London, 24–27 June; 2013. p. 239–48.

[18] Yang M, AlKutubi M, Pham D. Continuous acoustic source tracking for tangible acoustic interfaces. Measurement 2013;46(3):1272–8.

[19] Qin W, Suo Y, Shi Y. CAMPS: a middleware for providing context-aware services for smart space. International conference on grid and pervasive computing, Taichung, Taiwan, 3–5 May; 2006. p. 644–53.

[20] Castanedo F. A review of data fusion techniques. Sci World J 2013;1–19.

[21] Von Neumann J. Probabilistic logics and the synthesis of reliable organisms from unreliable components. In: Automata studies. 34. Princeton, NJ: Princeton University Press; 1956. p. 43–98.

[22] Chow C. Statistical independence and threshold functions. IEEE Trans Electron Comput 1965;14(1):66–8.

[23] Hashem S. In: Algorithms for optimal linear combinations of neural networks. International conference on neural networks, Houston, 12 June; 1997. p. 242–7.

[24] Varshney P. Distributed Bayesian detection: parallel fusion network. New York: Springer; 1997. p.215, ISBN: 978-1-4612-7333-2215.

[25] Breiman L. Stacked regressions. Mach Learn 1996;24(1):49–64.

[26] Juditsky A, Nemirovski A. Functional aggregation for nonparametric estimation. Ann Stat 2000;28(3):681–712.

[27] Granger C. Invited review combining forecast–twenty years later. J Forecast 1989;8(3):167–73.

[28] Jagadish H, Gehrke J, Labrinidis A, Papakonstantinou Y, Patel J, Ramakrishnan R, et al. Big data and its technical challenges. Commun ACM 2014;57(7):86–94.

[29] Wu X, Zhu X, Wu G, Ding W. Data mining with big data. IEEE Trans Knowl Data Eng 2014;26(1):97–107.

[30] Luo R, Kay M. Multisensor integration and fusion: issues and approaches. Proc SPIE 1988;(April 4):42–9.

[31] Luo R, Yih C, Su K. Multisensor fusion and integration: approaches, applications, and future research directions. IEEE Sensors J 2002;2(2):107–19.

[32] Thomopoulos S. Sensor integration and data fusion. J Field Rob 1990;7(3):337–72.

[32a] Luo R, Kay M. Multisensor integration and fusion: issues and approaches, In: Proc. technical symposium on optics, electro-optics, and sensors, 1988.

[33] Pau L. Sensor data fusion. J Intell Robot Syst 1988;1(2):103–16.

[34] Harris C, Bailey A, Dodd T. Multi-sensor data fusion in defense and aerospace. Aeronaut J 1998;102(1015):229–44.

[35] Bedworth M, O'Brien J. The omnibus model: a new model of data fusion? IEEE Aerosp Electron Syst Mag 2000;15(4):30–6.

[36] Esteban J, Starr A, Willets R, Hannah P, Bryanston-Cross P. A review of data fusion models and architectures: towards engineering guidelines. Neural Comput Appl 2005;14(4):273–81.

[37] Khaleghi B, Khamis A, Karray F, Razavi S. Multisensor data fusion: a review of the state-of-the-art. Inf Fusion 2013;14(1):28–44.

[38] Wellington S, Atkinson J, Sion R. In: Sensor validation and fusion using the nadaraya-watson statistical estimator. Proceedings of the fifth international conference on information fusion, Maryland; 2002. p. 321–6.

[39] Zadeh L. A simple view of the Dempster-Shafer theory of evidence and its implication for the rule of combination. AI Mag 1986;7(2):85.

[39a] Kumar M, Garg D. Multi-sensor data fusion in presence of uncertainty and inconsistency in data. In: Sensor and data fusion. InTechOpen; 2009.

[40] Zhang K, Li X, Zhu Y. Optimal update with out-of-sequence measurements. IEEE Trans Signal Process 2005;53(6):1992–2004.

[41] Liggins M, Hall D, Llinas J. Handbook of multisensor data fusion: theory and practice. Boca Raton, FL: CRC Press; 2008. p.870, ISBN: 978-1420053081.

[42] González W, Mejía J. Fusión de datos en redes de sensors: una revisión del estado del arte. J Polit 2015;10(19):135–45.

[43] Siciliano B, Khatib O. Springer handbook of robotics. Berlin: Springer; 2016. p.2227, ISBN: 978-3319325507.

[43a] Blackman S, Broida T. Multiple sensor data association and fusion in aerospace applications. J Robot Syst 1990;7(3):445–85.

[44] Mahadevan S. Monte Carlo simulation. Mechanical engineering. New York, Basel: Marcel Dekker; 1997123–46.

[45] Zadeh L. The role of fuzzy logic in the management of uncertainty in expert systems. Fuzzy Sets Syst 1983;11(1):199–227.

[46] Chen M, Mao S, Liu Y. Big data: a survey. Mob Netw Appl 2014;19(2):171–209.

[47] Anokhin P, Motro A. In: Data integration: inconsistency detection and resolution based on source properties. Proceedings of international workshop on foundations of models for information integration, Viterbo, Italy, September; 2001. p. 1–15.

[48] Wang X, Huang L, Xu X, Zhang Y, Chen J. A solution for data inconsistency in data integration. J Inf Sci Eng 2011;27(2):681–95.

[49] Kumar M, Garg D. Multi-sensor data fusion in presence of uncertainty and inconsistency in data. In: Sensor and data fusion. Vienna: Intech; 2009. p.436, ISBN: 978-3-902613-52-3 p.436.

[50] Ssu K, Chou C, Jiau H, Hu W. Detection and diagnosis of data inconsistency failures in wireless sensor networks. Comput Netw 2006;50(9):1247–60.

FURTHER READING

[1] Castanedo F. Fusión de datos distribuida en redes de sensores visuales utilizando sistemas multi-agente, Available from, https://e-archivo.uc3m.es/handle/10016/9495; 2010. Accessed September 2016.

WIRELESS SENSOR TECHNOLOGY FOR INTELLIGENT DATA SENSING: RESEARCH TRENDS AND CHALLENGES

3

Djallel Eddine-Boubiche*, Joel A. Trejo-Sánchez[†], Homero Toral-Cruz[‡], José L. López-Martínez[§], Faouzi Hidoussi[¶]

University of Batna 2, Batna, Algeria* CONACYT-Center for Research in Mathematics, Merida, Mexico[†] Center for Research in Mathematics, University of Quintana Roo, Chetumal, Mexico[‡] University of Yucatan-Department of Mathematics, Merida, Mexico[§] Corgascience Company, Borehamwood, United Kingdom[¶]

3.1 INTRODUCTION

Wireless sensor technology has been recognized as one of the emerging technologies of this century, widely used for intelligent data sensing. A wireless sensor network (WSN) is composed of several sensor nodes, where the main objective of a sensor node is to collect information from its surrounding environment and transmit it to one or more points of centralized control, called base stations or sinks, for further analysis and processing [1]. A sensor node is limited in resources and capabilities and mainly has two components: the component responsible for storage, computation, and communication, called the node; and the component responsible for intelligent data sensing, called the sensor. Due to the previously mentioned limitations, the design and implementation of a WSN has several challenges, such as power-efficient management, efficient communication protocols, bandwidth, storage, processing, and security issues.

As a result of the constraints imposed on the sensor nodes, many researchers have focused on the design of efficient protocols, techniques, and methods for sensing, transmitting, and analyzing data using wireless sensor technology.

Data aggregation is an important basic concept in WSNs, proposed to optimize the data collection process and the sensor nodes' energy reserves. Indeed, due to the network density, the data collection process suffers from redundancy and data interrelation, which may drain the sensor node batteries and affect the network performance (overhead, transmission bandwidth, latency). Thus it is necessary to use methods for merging data at the intermediate nodes to reduce the number of packets transmitted to the base station, allowing energy and bandwidth savings. This can be accomplished through a data aggregation mechanism. In this chapter we present a survey of the existing security aggregation protocols dedicated to heterogeneous WSNs.

Intelligent Data Sensing and Processing for Health and Well-being Applications. https://doi.org/10.1016/B978-0-12-812130-6.00003-2

On the other hand, routing protocols are important issues for those networks where resources are limited. In this chapter we present an overview of the main routing protocols and describe the advantages and disadvantages of these protocols. We first present some protocols based on geographic coordinates. These include MFR [2], facing routing protocols [3], and others. Later, we present nongeographic-based routing protocols, for instance those using position trees [4]. Next, we present some hierarchical or cluster-based routing protocols, which are well-known techniques with special advantages related to scalability and efficient communication. Within these last protocols we present the LEACH [5], TEEN [6], and others. Finally, we discuss some routing protocols for WPAN and WBAN networks.

Finally, we describe some strategies that would be useful to incorporate for sensor node mobility. We explain the self-stabilizing property of WSNs. Additionally we explain the importance of messages of short size in WSNs and describe some trends in selecting a high centrality leader for the coordination protocols in WSNs.

The main contribution of this chapter is to present an overview of the main research trends and challenges of wireless sensor technology, in order to understand intelligent data sensing and processing for health and well-being applications.

3.2 DATA AGGREGATION IN WIRELESS SENSOR NETWORKS

The main function of a sensor node is to collect measurements of the environment in which it is deployed and to collaborate with other nodes for routing the data to a processing center, called the base station. These sensors are generally limited in energy and have a reduced storage capacity. Since the sensor nodes are constrained in energy, it is inefficient for all the sensors to transmit data directly to the base station, so the multihop routing protocol is always applied to relay data to the base station. Indeed, data generated by neighboring sensors are often redundant and highly correlated. In addition, the data amount generated in a large-scale sensor network may exceed the processing capacity of the base station. Consequently, it is necessary to use mechanisms to merge and aggregate data at intermediate nodes (relay nodes) in order to reduce the number of packets transmitted to the base station, thereby saving energy and bandwidth. This can be accomplished through a data aggregation mechanism.

Data aggregation is defined as the process of merging data from multiple sensors at intermediate nodes in order to eliminate redundant transmissions. Data aggregation aims to collect the most crucial data from the sensor nodes and make them available to the base station in an energy-saving manner and with minimal data latency. Indeed, optimizing data latency is important in many applications, such as environmental monitoring where freshness of data is a major factor.

Data aggregation presents a number of advantages, which can be summarized as follows:

- By eliminating redundancies, data aggregation can improve the robustness and precision of the obtained information from the network,
- Data aggregation reduces traffic, which contributes to energy conservation of sensor nodes.

However, data aggregation may also have some disadvantages:

- The nodes responsible for the aggregation are subject to attacks. If such nodes are compromised, the base station cannot verify the integrity of the aggregated received information.
- Multiple copies of aggregated information may be sent to the base station, which may increase energy consumption.

3.2.1 **DATA AGGREGATION PERFORMANCE METRICS**

Critical performance metrics for data aggregation algorithms are closely related to the application fields: energy efficiency, network lifetime, latency, and data accuracy. Indeed, the data aggregation process should significantly reduce the overall network work load and thereby optimize energy efficiency and the network lifetime. The accuracy represents an important evaluation metric in the data aggregation process. Several definitions of data aggregation accuracy are used, the most common being the number of collected data points received by the base station (N_{agg}) divided by the number of data points sensed and sent from the network sensor nodes (N_{sens}): Accuracy $= N_{agg}/N_{sens}$.

The data aggregation accuracy depends on the introduced end-to-end delay. Indeed, for better data aggregation accuracy the aggregator nodes must wait for a longer time to receive collected data from all their neighboring nodes, which leads to extra end-to-end delay. A good aggregation algorithm must determine the required data aggregation amount in order to meet time constraints and maintain acceptable latency bounds on data delivery with minimum energy consumption.

3.2.2 **DATA AGGREGATION APPROACHES**

Based on the network architecture adopted, we can classify the data aggregation approaches into three categories:

- *Central aggregation*: based on a cluster-based network architecture formed through a clustering protocol [7]. The network is divided into clusters and in each cluster a leader node (cluster head) is elected to manage all nodes in the cluster and aggregate collected data. This cluster head may possibly change over time in order to distribute the energy consumption as much as possible between all the nodes of the cluster (Fig. 3.1).
- *Distributed aggregation*: the aggregation process is handled through a network tree architecture [8]. The base station represents the root and the nodes represent the leaves. The information starts from the leaves until arriving at the root, and redundant data are aggregated at each tree level. The disadvantage of this approach is that the networks are not free to fail. If data packets are lost at any level in the tree, the data will be lost, not only for a single level but for the entire associated subtree (Fig. 3.2).
- *Hybrid aggregation*: combines the characteristics of the previous two types of aggregation. The hybrid aggregation scheme can dynamically change from one aggregation technique to the other in an unpredictable environment and adapt to dynamic changes in the network [9]. Therefore the hybrid scheme would take the best of both central and distributed approaches.

The data aggregation approach can be also classified based on the network composition.

- *Data aggregation in homogeneous WSNs*: In homogeneous networks, all nodes are identical and have the same roles. In such networks, data aggregation is performed based on datacentric routing protocols.
- *Data aggregation in heterogeneous WSNs*: Data aggregation in heterogeneous WSNs involves merging data at particular nodes, which reduces the number of messages transmitted to the base station and improves the network's energy efficiency.

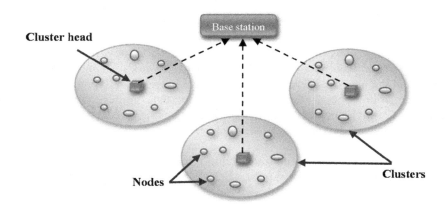

FIG. 3.1

Central aggregation.

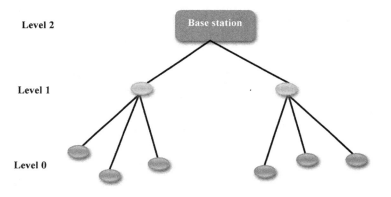

FIG. 3.2

Distributed aggregation.

3.2.3 DATA AGGREGATION TECHNIQUES

The data aggregation fusion and merging is based on two different techniques, called lossy and lossless aggregation.

- *Lossless aggregation*: In lossless aggregation, data packets are concatenated into larger packets with no data loss. Indeed, lossless aggregation is suitable as long as the workload is less than the network capacity.
- *Lossy aggregation*: An example of this aggregation type is reporting the sensed data average instead of all sensed data. Lossy aggregation is suitable when the network capacity is limited, to preserve sensor node resources.

3.2.4 DATA AGGREGATION PROTOCOLS

Since data aggregation is done at the routing level, most routing protocols dedicated to WSNs adopt this principle to improve network performance. Thus, routing protocols can be classified according to the applied data aggregation approaches. Fig. 3.3 illustrates some known WSN routing protocols grouped by data aggregation approaches.

3.2.5 DATA AGGREGATION SECURITY

The data aggregation process is relatively trivial, but becomes difficult when one wants to add security and especially confidentiality (encryption). In some applications, it is essential to ensure that information transmitted over the network cannot be intercepted and read by unauthorized persons. According to Ref. [10], a secure data aggregation process can be divided into three phases:

- *Requests dissemination phase*: The base station broadcasts requests throughout the network.
  Example:

SELECT AVERAGE (temperature) FROM Sensors.
WHERE floor=6

- *Data aggregation phase*: All nodes satisfying the base station query are referred to as source nodes and send their captured data. All nonleaf nodes in the routing tree aggregate the data received from their down tree level using an application-specific aggregation function.
- *Verification phase*: The base station checks the aggregation result with all source nodes. Another designation for this mechanism is attestation, as nodes must attest to the validity of the aggregation result. Given the complexity of communication, this phase is reserved only for aggregation results in the case of critical missions. Thus, this phase is not always executed (Fig. 3.4).

3.2.5.1 Basic requirements for data aggregation security
- *Data confidentiality*: preserving the data confidentiality during the data aggregation process can be implemented in two ways: hop-by-hop and end-to-end. In the hop-by-hop approach, each

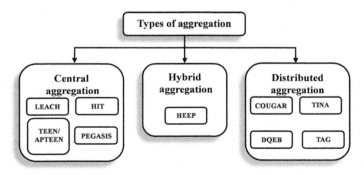

FIG. 3.3

Data aggregation protocols.

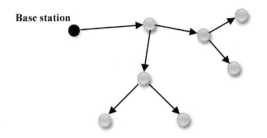

FIG. 3.4

Phase of requests dissemination.

aggregation node must decrypt the received data, apply the aggregate function and reencrypt the data before sending it to the next aggregation node. This type of privacy implementation requires additional computation, which takes more time and consumes more energy. However, in the end-to-end approach, there is no need to decrypt and reencrypt the data at each aggregation node. The aggregation functions are directly applied to the encrypted data using specific mechanisms, such as homomorphic encryption.

- *Data integrity*: Even if the network can guarantee the data confidentiality, integrity may still be affected. In the case where the attacker can compromise the aggregation node, the aggregation security mechanisms must guarantee that the attacker cannot modify the aggregated data.
- *Data freshness*: When an attacker intercepts a message, it could reuse it later. Freshness of data must be ensured, especially in aggregation protocols based on key management mechanisms.
- *Authentication*: The data aggregation security mechanisms must ensure that an adversary node cannot claim to be an aggregation node and falsify the aggregation results.

3.2.5.2 Data aggregation attacker profile

Data aggregation security mechanisms are subject to different types of threats, which may come from different types of attackers. An attacker can access one or more sensor nodes, thus compromising them and modifying any information contained therein. In this case the attacker can be either passive [11], with the role of spying on traffic in order to obtain important information about the data, or active, thus having the power to destroy the data, modify them, reach their destination, or compromise the nodes. Another scenario is that in which an attacker has access to all the nodes of the network. A passive attacker can listen to all network traffic, and an active attacker can access and compromise all network nodes (sensors, aggregation nodes, base station). Based on these two scenarios, active attackers can be categorized as strong attackers with full access to the network and thus able to launch different attacks against any node on the network, or they may be weak, able to spy on parts of the network but not able to interact with the whole of the network.

3.2.5.3 Data aggregation security attacks

In the following text, we will present attacks on the WSNs that are able to directly affect the aggregation mechanisms of the data.

- *Sybil attack* [12]: In this type of attack, the attacker is able to present several identities. A Sybil attacker node can affect the data aggregation mechanism in different ways:

- It can create several identities in order to generate additional votes in the election phase of the aggregation nodes in order to be able to inject a malicious node.
- It can also generate multiple input nodes with invalid data in the aggregate function.
- *Selective forwarding attack* [13]: Each node that receives a message must retransmit it. However, a compromised node may refuse to transmit all of the received data or some of it, which is called selective forwarding. In the data aggregation process, a malicious aggregation node can select the data to be aggregated or prevent the transmission of some aggregated data.
- *Replay attack* [13]: In this type of attack, the attacker can spy on traffic and reuse old messages. Thus it can retransmit old aggregation results, falsifying the new results accordingly. This type of attack is the easiest to implement and the most difficult to be detected, as long as the malicious node does not need to access the internal memory of the nodes or intercept the encrypted data.
- *Spoofed data attack* [13]: In this type of attack, the malicious node intercepts and modifies the data by injecting erroneous values into it. The node can thus intercept the aggregated data and modify it. This attack does not work alone; the malicious node must combine it with either the replay or selective forwarding attack.
- *Sinkhole attack* [14]: In this attack the attacker tries to make the compromised node very attractive to its neighboring nodes so that neighboring nodes will choose it as the next node in their routing path in order to route their data. It can thus modify the data already aggregated or awaiting aggregation. The compromised node can thus modify, delete, or even reject the received packets, which considerably affects the proper functioning of the network.
- *False data injection attack* [15]: The attacker aims to inject false data into the network without revealing its existence. In a data aggregation scenario, the value of the injected data leads to a false aggregation result. A compromised node can report significantly biased or fictitious values and perform a Sybil attack, thereby affecting the aggregation result.

3.2.5.4 *Data aggregation security protocols for heterogeneous WSNs*

Several security primitives are widely used in securing data aggregation in WSNs. These primitives include ECC (elliptic curve cryptography), privacy homomorphism, digital signatures, and hash function. These primitives form the basis of several data aggregation security protocols. A large number of data aggregation protocols have been proposed to secure the data aggregation process in classical homogeneous WSNs. However, homogeneous WSNs impose several limits on the applied data aggregation security protocols due to the simple sensor nodes resource capacity. Therefore heterogeneous WSNs offer better research fields in the context of data aggregation security, where complex and high-level security primitives can be applied, such as privacy homomorphism. Only a few data aggregation security protocols have been proposed in heterogeneous WSNs, with SRDA (Secure Reference-based Data Aggregation protocol for wireless sensor networks, CDAP (Concealed Data Aggregation in Heterogeneous Sensor Networks using Privacy Homomorphism), PIA (Privacy-preserving Integrity-assured data Aggregation), SDAKM (Secure Data Aggregation with Key Management), and RCDA (Recoverable Concealed Data Aggregation for data integrity in wireless sensor networks) being the best known.

3.2.5.4.1 SRDA

Sanli et al. [16] proposed a secure hop-by-hop data aggregation system that allows encryption of the aggregation result and applies varying levels of security across the hierarchy (aggregation nodes). The proposed protocol is based on the principle that the level of security should be gradually increased as messages are transmitted into higher network levels. Therefore, a cryptographic algorithm is introduced to adjust security parameters and the number of encryption cycles, so as to be able to change the security level if necessary. The basic idea behind SRDA is that the nodes transmit referential data instead of the pure captured data and only the difference between these two types of data is transmitted. According to this approach, the encrypted differential data representing the difference between the reference value and the sensed data are transmitted to the aggregation points in place of the sensed data in order to reduce the number of bits transmitted. SRDA first provides a key distribution scheme with low memory overhead to establish secure communication links in the network. Then, to save energy, SRDA applies a variable security-level mechanism at a different network hierarchy.

SRDA not only reduces the number of bits transmitted and increases network security, but it also reduces power consumption and bandwidth by no longer sending the collected data. However, the authors did not give any indication of the type of network on which the proposed protocol can be used. Nevertheless, given the fact that the mechanisms of aggregation and security occur hierarchically at the level of specific nodes, it can be assumed that these nodes must be superior to the sensor nodes in terms of computing power and energy. In the context of a heterogeneous WSN, the task of aggregation and security must go back to the heterogeneous nodes so that the operation of the network is optimal.

3.2.5.4.2 CDAP

An approach proposed by Ozdemir [17] to facilitate the implementation of secure aggregation and communication where the difficulty arises from the need to aggregate encrypted data, CDAP proposes to use a privacy homomorphism that offers end-to-end data concealment and the ability to operate on cryptograms, while aggregating the data.

Indeed, the homomorphism can be based on symmetric or asymmetric key cryptography. However, the use of symmetric keys may be vulnerable to attacks. Thus, the author proposes to use the privacy homomorphism based on asymmetric keys. Given that the asymmetric cryptographic confidentiality homomorphism incurs high computational costs for encryption and aggregation, it cannot be guaranteed by nodes with limited resources. To this end, the approach uses a set of resource-rich nodes, called aggregation nodes (AGGNODEs), for encryption based on privacy homomorphism and aggregation of encrypted data.

CDAP increases the ability to secure the data aggregation mechanism in the presence of compromised nodes. The author evaluates the performance of the network using AGGNODEs compared to a homogeneous WSN, where only the base station represents superior equipment in terms of computing capacity and memory. The evaluation results demonstrate CDAP improvements on data aggregation security, energy efficiency, and bandwidth used. The performance evaluation also shows that the protocol can be applied over large heterogeneous WSNs.

3.2.5.4.3 PIA

In this approach proposed by Taban et al. [18], the authors highlight the problem of data aggregation integrity and privacy as a common goal. PIA assumes a network model where the aggregation node is used as an intermediary between the user and the sensor nodes. The main problem is that the user wants to verify the integrity of the aggregated data received, while the owner of the network wants this data to remain secret for the user. To do this, the authors propose four schemes. The first scheme uses the privacy homomorphism to hide the data. The homomorphism combined with the MAC authentication message is used to construct an authenticated encryption scheme for the aggregation model. The second scheme uses the OPES approach [19], which preserves the privacy of data dissemination. The third scheme uses the SHIA approach [20] to adapt distributed integrity checking; the main characteristic of this scheme consists of supporting any aggregation function. However, the communication cost of this scheme is O(N) messages per sensor node, where N represents the total number of nodes in the network. To ensure the privacy and integrity of the data, the third scheme is improved by the introduction of an aggregation logic tree within the aggregation node. This represents the last schema in which the communication cost becomes O ($\log N$). However, this scheme only supports a few aggregation functions, such as average and min/max.

PIA offers a good data aggregation security level; however, the authors give no indication of the deployment of the heterogeneous nodes and of the contribution that the heterogeneity can make within the framework of this protocol.

3.2.5.4.4 Secure encrypted-data aggregation for wireless sensor networks

In this protocol proposed by Huang et al. [21], the authors propose a secure data aggregation protocol that aims to eliminate redundant sensor readings without encryption and to maintain confidentiality of the data during transmission. The proposed scheme ensures the security and confidentiality of the data, and the redundant data is aggregated into a single packet.

The proposed protocol is dedicated to centralized WSNs and the network is divided into groups of nodes with a leader in each group that performs the aggregation function. Group leaders have stronger radio antennas that can transmit directly to the base station. The approach assumes that the sensor nodes can only transmit to the aggregation nodes, thus reducing transmission costs and energy power. The authors propose an aggregation mechanism that can maintain the confidentiality and privacy of the data. Thus, the sensor nodes encrypt the data before transmitting them to the aggregation nodes, but these data remain secret to the aggregation node, which cannot know anything about them. In addition, the approach can also prevent known plain-text attacks and selected text attacks.

Indeed, the aggregation nodes do not need to decrypt the data received from the sensor nodes before applying the aggregation function, thus saving more energy. The data is encrypted using XOR operators and a hash function. The use of random keys for data encryption makes the approach more robust against attacks of malicious nodes. The approach offers more confidentiality since each node can randomly generate a new key, thus ensuring network security. However, it is believed that the deployment of heterogeneous nodes is not optimal, as long as data encryption is done at energy-limited sensor nodes and by using simple encryption algorithms.

3.2.5.4.5 SDAKM

Sandhya and Murugan proposed [22] the SDAKM protocol, the main objective of which is to provide a secure scheme for the given heterogeneous WSNs using the privacy homomorphism. SDAKM uses the heterogeneity of sensor nodes to perform encrypted data processing and also provides an effective key management system for communicating data between the sensor nodes in the network.

SDAKM assumes the deployment of aggregation nodes with more computational capacities than the simple sensor nodes. The authors suppose that the simple sensor nodes transmit encrypted data to the aggregation nodes. Therefore, aggregation nodes must be able to perform the processing on the encrypted data. SDAKM performs a secure aggregation of the data using the Hill Cipher (IHC) iteration technique. It consists of an additive confidentiality homomorphism system for the communication between aggregation nodes. As mentioned, an efficient key management system is also introduced to secure communication between the sensor nodes in the network. This key management system is based on key preallocation and revocation using a key management decentralized approach. The authors propose to use such an approach to solve the revoking keys problem in compromised nodes.

SDAKM significantly increases the security level of the network by providing a secure scheme through a data encryption process for aggregation and efficient and decentralized key management. Indeed, to decipher an IHC encrypted text, a secret key matrix A and a number of iterations k must be known. Therefore, even if an attacker knew the A matrix, without knowledge of k, he would not be able to decrypt any given encrypted text. Therefore, if the number of iterations k is kept secret as part of the private key, the security of IHC is increased.

3.2.5.4.6 RCDA

Recoverable Concealed Data Aggregation, or RCDA, was proposed by Chen et al. [23] for data integrity in WSNs. In RCDA, the base station can retrieve all the data collected even if the data has been aggregated, in order to confirm the integrity and authenticity of the data (in the other protocols, the base station cannot recover aggregated data). The proposed approach is adopted for both homogeneous and heterogeneous WSNs. The authors propose two schemes dedicated to homogeneous and heterogeneous RCSFs: RCDA-HOMO and RCDA-HETE. We will be interested in RCDA-HETE. In RCDA-HETE, there are two types of nodes: L-nodes (low end), which make up the majority of the nodes of the network and which have low capacity, and H-nodes (high end), consisting of more powerful nodes and which represent resistant nodes to attacks, allowed to store the keys if necessary.

RCDA-HETE consists of five procedures: installation, intragroup encryption, intergroup encryption, aggregation, and verification.

- During the installation process, the necessary keys are loaded into the L-nodes and H-nodes. Each L-node must share a key with its group leader. If the base station knows the group information before deployment, the keys can be preloaded on all L-nodes and H-nodes. However, in most WSNs, the nodes are deployed in a random manner.
- The intragroup encryption procedure is performed when the L-nodes wish to send their captured data to the corresponding H-node. It aims to establish a secure channel between the nodes. For example, a node Li encrypts the data with its key Ki and sends them to a node H1 which decrypts them.
- In the intergroup encryption procedure, each H-node aggregates the received data and then encrypts and signs the result. In addition, if an H-node receives encrypted data and signatures from other nodes in its routing path, it activates the aggregation function.
- Finally, the verification procedure carried out by the base station ensures the integrity and the authentication of the aggregation results.

RCDA-HETE ensures the integrity and the authentication of the data. Indeed, RCDA-HETE allows checking each piece of data using the H-nodes. Specifically, the intergroup encryption procedure allows the L-nodes to send not only the encrypted data to their head, but also a MAC authentication message, which will allow the L-nodes to verify the integrity of the data.

RCDA-HETE also makes it possible to use a multitude of aggregation functions preloaded into the H-nodes before the network is deployed. Thus, the base station may request these nodes to perform an aggregation function designated according to the needs of the application.

In RCDA-HETE, the intragroup traffic is encrypted by a pair of keys and the protocol generates the encryption signature corresponding to each data point. As a result, attacker nodes cannot modify messages or inject false data as long as they cannot alter messages without private keys. If a malicious node compromises the sensor nodes, the following situations are to be considered. The adversary can compromise a sensor node and use it as a legal node. Also, if the value of an altered message is within a reasonable interval, its detection remains difficult. A malicious node may also try to manipulate the aggregation results by generating false data, changing legitimate messages, or impersonating other nodes. RCDA-HETE provides security against these types of attacks by using a signature for each generated message.

3.2.5.4.7 A watermarking-based mechanism for data aggregation security in heterogeneous WSNs

Boubiche et al. proposed a two-level data aggregation security protocol, which applies the watermarking concept and exploits the benefits of the cross-layer architecture approach. The main idea of this protocol is to apply an efficient, lightweight security scheme on resource-limited homogeneous simple sensor nodes, while the heterogeneous node capacity is exploited by using a high security level mechanism without restricting the use of data aggregation functions. Two algorithms have been proposed at each network level to ensure data aggregation security according to the sensor node resource capacities. Indeed, the "Secure Data Aggregation Watermarking-based Scheme in Homogeneous WSNs" (SDAW) algorithm was proposed at the low level of the network, and the "Cross-layer Watermarking-based Mechanism for Data Aggregation Integrity in Heterogeneous WSNs" (CLDWA) algorithm was introduced at the high level of the network [24].

The SDAW algorithm is mainly based on the generation of a lightweight watermark using the collected data combined with the sensor node identifier (using a simple XOR function). A one-way hash function SHA-1 is applied to the obtained result to generate the final watermark. The watermark is then concealed in the data packet so that it will represent an integrated part of the original data, which is also inserted into the data packet. When the data packet is received, the receiver node will extract the watermark. Then the node will use the original data contained in the data packet to generate a new watermark. The values of the extracted and the newly generated watermarks are compared. If they are identical, the node will accept the data, and if they are not identical the node will suppose that the data has been altered and will reject it.

At the high level of the network, the authors proposed to reinforce the fragile watermarking mechanism introduced by the SDAW algorithm with asymmetric encryption, where keys are used to generate and extract the watermark. Thus, the security between the heterogeneous nodes and the base station is enhanced while optimally exploiting the heterogeneous nodes' higher capacities. In addition, unlike the privacy homomorphism encryption, which offers restricted aggregation functions (only addition and multiplication functions are allowed), the proposed CLWDA algorithm does not apply any restrictions on the data aggregation functions [24].

3.3 ROUTING IN WIRELESS SENSOR NETWORKS

Routing is one of the most representative challenges in WSN, due to the amount of information that is sent in the WSN and the restriction in power of the sensors. The main task of a routing algorithm is the determination of optimal routes of transmission of the packets from the nodes to their destination [25]. Routing protocols may differ depending on the architecture of the network and its applications. Routing protocols for WSNs face a number of challenges, mainly due to the fact that WSNs are very different from traditional communication networks [26]. We describe some of these challenges. The *node deployment* consists of the physical location of each of the sensors in the WSN. The node can be positioned manually in its respective location or can be placed randomly, creating a network topology. Clustering algorithms are necessary if the distribution of the nodes is not uniform. The *energy consumption* constitutes a challenge, since the sensors in WSNs are energy restricted; therefore the routing protocol must be efficient in its communication. The *node heterogeneity* assumes that the nodes can have different roles in the WSN. The routing protocols shall consider the diverse profiles of the sensors to allow an efficient node heterogeneity communication. The routing protocols should consider adjustment transmit powers and signal rates when a *subset of sensors fails* due to an arbitrary fault condition. The *connectivity* refers to the degree of communication of a sensor. The sensors in a WSN are expected to be highly available for communication. This degree of communication depends generally on the network topology and the distribution of the nodes.

Considering these challenges, the routing protocols can be divided [27] into flat-based, hierarchical-based, and location-based routing protocols, if we are aware of the network structure. In the *flat-based* protocol, all the sensors behave similarly; in the *hierarchical-based* protocol, the sensor performs different roles in the WSN; and in the *location-based* protocol, the routing considers the location of the sensor in the WSN related to routing in the network.

The routing protocols can be classified into multipath-based, query-based, negotiation-based, QoS-based, and coherent-based routing protocols if we are aware of the protocol operation. The *multipath-based* protocols consider diverse paths to route the data in the network. In *query-based* routing protocols, the sensors communicate through queries. A sensor x sends a query to its neighbor y, and the sensor y replies with the result of such a query. In *negotiation-based* routing protocols, the communication is through data descriptors, with the objective being to eliminate the redundancy in messages by using negotiation mechanisms. In QoS-based protocols, the communications are aware of maintaining a balance between the consumption of energy and quality of the data. In coherent-based protocols, the sensors cooperate to process different data in the communication. Table 3.1 provides a summary of the classification of the routing protocols proposed by Al-Karaki et al. The reader can refer to Ref. [27] for a detailed description of these classifications.

Table 3.1 Classification of Routing Protocols According to Al-Karaki et al.

Routing Protocols		
Network structure	Flat-based	Sensors have similar behavior
	Hierarchical-based	Sensors perform different roles
	Location-based	Sensors consider their position in the WSN
Protocol operation	Multipath-based	Multiple paths
	Query-based	A sensor sends a query and receives the reply from the query
	Negotiation-based	Communication through data descriptors
	QoS-based	Balance between energy consumption and quality of the data
	Coherent-based	Sensors cooperate to process different data

In sensor networks, it is necessary to transmit the flow of information from multiple sources to a sink, unlike the architectures of traditional distributed systems. Another feature is the redundancy of information transmitted by sensors in a neighborhood being affected by a phenomenon in common. According to the WSN distribution of sensors, the routing protocol can be flooding or nonflooding. In flooding protocols, when a sensor receives a message, it broadcasts the message to all its neighbors. Fig. 3.5 shows the behavior of the flooding protocol.

Nonflooding protocols are efficient since they consume lower energy. Location-based protocols are preferable since these protocols improve the efficiency of the routing protocols. Next, we describe some of these protocols.

The most forward routing. The behavior of the most forward routing (MFR) [2] is as follows. The requested sensor sends the message to the sensor that is closest to the sink. In this protocol, it is assumed that each sensor knows its location; this is a valid assumption since the sensor can have GPS. This protocol performs well if the WSN is dense; however, if the WSN is sparse, the message can possibly never be delivered to the sink, if trapped in a local set of nodes. In Fig. 3.6 the message is not delivered to the sink. The message traverses from node x to node y indefinitely.

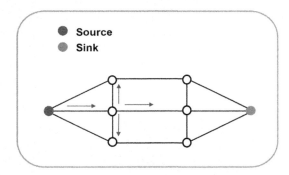

FIG. 3.5

Flooding protocol. When a sensor receives a message, it broadcasts to all of its neighbors.

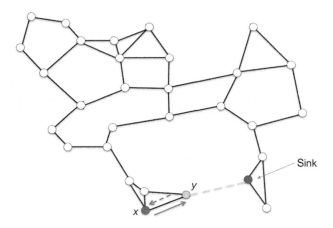

FIG. 3.6

Message trapped in a set of nodes in the MFR protocol.

The face routing protocol is a mechanism to avoid the messages to be trapped on a set of sensors.

Face routing protocol. The face routing protocol [3] ensures the construction of a route if there exists a route in a planar WSN. The face routing protocol constructs a route from the source sensor to the sink by exploring the faces of the planar graph, considering a destination line from the source to the sink. The face routing protocol generates the path by exploring the planar WSN face by face. There exists some greedy strategies detailed in Ref. [3] for some WSNs that can be trapped in a local minimum, particularly for the case of sparse graphs.

Virtual coordinate assignment protocol. The last two protocols require the use of a coordinate system, such as a GPS, to generate a valid routing protocol; however, assuming that every sensor has a GPS system is not always feasible. Some authors [28] consider the use of a virtual coordinate system to estimate approximate values of the coordinates of all the sensors in the WSN. Unfortunately, the virtual coordinate system of Ref. [28] requires that initially a subset of cardinality proportional to $O(n)$ contain a traditional coordinate system. Then, the same cost restriction as the location routing protocols is also in protocols of virtual coordinates.

Position trees protocol. The routing position tree protocol [4] generates a virtual coordinate system in a WSN. The position trees allow the creation of a virtual label for each one of the sensors such that each sensor uses low memory. This protocol allows definition of virtual coordinates, constructing a spanning tree rooted at an arbitrary sensor r. When the spanning tree is generated, it is possible to send a packet from a source x to a sink sensor y using the least common ancestor. Fig. 3.7 shows a WSN with a label generated with the position trees protocol. To send a package from the target r1111 to the sink

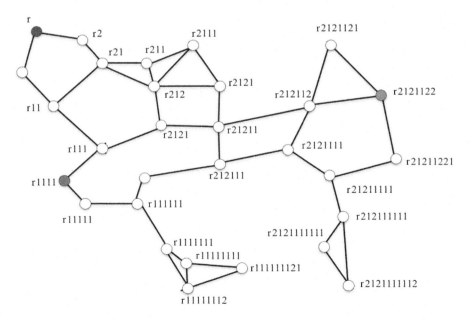

FIG. 3.7

Position tree. A package from r1111 to r2121122 traverses the path r1111-r111-r2121-r212-r2121-r21211-r212112-r2121122.

r2121122, the package traverses a path from the target to the lowest ancestor r212. Then the package traverses from the lowest ancestor r212 to the sink r2121122.

In addition to ensuring packet delivery, one of the most important challenges in routing is fault tolerance. Next we describe the main strategies for fault tolerance in sensor networks.

3.4 FAULT TOLERANCE IN WSN

Since the WSN performs similarly in a distributed fashion, we first introduce formally a distributed system. A *distributed system* can be modeled as a graph $G = (V, E)$, where each node $v \in V$ represents a sensor. Each sensor x communicates with a subset of sensors; we refer to such subsets as neighbors, $N(x)$, and the communication between any pair of neighbors is represented by an edge $e \in E$. We refer to the subset $N(x)$ as the open neighborhood, whereas the closed neighborhood is the subset $N[v] = \{N(v) \cup v\}$. Generally, WSNs are modeled with unit disk graphs. A unit disk graph is associated with a disk in the plane. Each vertex x in the UDG is the center of a unit disk. There exists an edge between vertices x and y if and only if the distance from x to y is at most one unit. Fig. 3.8 represents a unit disk graph and, as you can see, each vertex has a radius of influence. In a unit disk graph two vertices x and y are neighbors if their radius of interference intersects.

Neighbors in a distributed system communicate through message passing or by shared memory [29]. The transient faults consist of a corruption in the state of a single node, but not in its behavior. One efficient technique to make a system tolerant to transient faults is self-stabilization.

A system is self-stabilizing when, independently of its initial configuration, it converges to a legitimate state in a finite number of steps. Thus a self-stabilizing system does not require any initializing of its variables. A self-stabilizing system is capable of recovering from transient failures (such failures that corrupt the state of the nodes, but do not interfere with its behavior). Some WSN self-stabilizing protocols exist. Fig. 3.9 illustrates the concept of self-stabilization. The first state of a self-stabilizing system is arbitrary, but after a finite number of step the system converges to a legitimate state. Once the system is in a legitimate state, any transition yields a legitimate state.

Self-stabilization was first defined by Dijkstra in his seminal work [30], as a system that, independently of its initial state, converges to a legitimate state in a finite number of steps. Self-stabilizing systems are initially defined over a distributed systems, but we extend the definition of self-stabilizing systems to WSN. A self-stabilizing system is composed of several sensors; each sensor has a local state, i.e., the values of the variables of such a sensor. The global state of the WSN is the union of

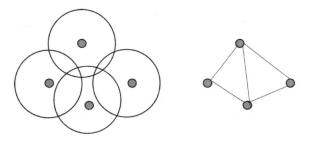

FIG. 3.8

Unit disk graph.

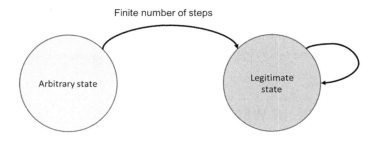

FIG. 3.9

Transition state of a self-stabilizing system.

the local states of each sensor. A self-stabilizing system can be defined if it satisfies the following two properties:

— Convergence. Independently of the initial state of a system S, the system converges to a legitimate state P in a finite number of steps.
— Closure. Once the system S is in a legitimate state P, it will remain in P.

The main challenge in defining a self-stabilization protocol is to ensure that, after a finite number of steps, the system converges to a legitimate state independently of its initial configuration. Now, we describe a self-stabilizing algorithm embedded in a WSN. The self-stabilizing algorithm consists of a set of rules in a sensor x of the form: $<$program$>$: guard $\rightarrow$ action. The guard is a Boolean expression of the variables in $N[x]$. The execution of an action updates the variables of x. If the guard of a rule is true, the sensor is enabled. If two or more sensors are enabled, a scheduler selects a subset of the sensors to execute the rule.

We now discuss some self-stabilizing algorithms defined over a WSN. We recommend two algorithms for clustering in a WSN.

Clustering is a mechanism to achieve scalable and efficient control over a WSN and save energy, by reducing the distance over which the sensors must send information through the WSN. Clustering consists of partioning the sensors into groups, referred to as clusters. In each cluster is a special sensor called the clusterhead, whereas the rest of the sensors are ordinary sensors. Only the leaders of each cluster must communicate with the base station. Routing protocols based in clustering reduces the amount of information that is propagated to the WSN. One of the major challenges in WSN clustering, is producing approximately equal diameter clusters, and minimize the overlapping among sensors in different clusters.

Demirbas et al. [31] designed a protocol for clustering in a WSN with the solid-disc property. The solid-disk property states that all the sensors at unit distance of the clusterhead i belong to the ith cluster. This property ensures that there exists no overlapping with sensors with unit distance. Furthermore, the cluster contains sensors at m-distance, where $m > 1$. Their protocol is self-stabilizing and scalable, which implies that it is tolerant to faults and changes in the network. Even more, their protocol has a constant stabilization time.

Later, Johnen and Nguyen [32] presented a robust self-stabilizing algorithm for clustering in a WSN. The clustering has the following properties: each ordinary sensor (not the cluster-head) is at distance 1 from the cluster-head of its cluster; in a neighborhood of a sensor x there are at most k cluster-heads. Each sensor selects among all the possible cluster-heads the best choice. Their algorithm converges after $O(D)$ asynchronous rounds.

With respect to fault tolerance in WSN, in Ref. [33] Miyaji and Omote present self-healing as a property to increase the availability of sensors in a WSN even if a subset of the sensors is attacked. The self-healing implies that the corrupted links are automatically repaired after a finite number of steps, even if the attacker compromises the sensor nodes of the network.

3.5 CONCLUSION

This chapter is focused on describing the challenges and trends in WSNs. First we provide a general classification of data aggregation protocols. We classify these protocols into three categories: central aggregation, distributed aggregation, and hybrid aggregation. Then, we provide in detail the main requirements for data aggregation security, considering the WSN architecture. We also describe the main challenges to routing in WSN and survey the most important protocols proposed in the field. The self-stabilizing concept has also been addressed in this chapter, where we present the main challenges in creating self-stabilizing protocols for WSN.

Indeed, efficient routing, self-stabilizing, and optimized data aggregation represent important prerequisites to build robust sensor networks. These three concepts have been widely studied and still attract the attention of researchers. The continued development of sensor technologies and the emergence of new application domains, such as underwater and underground WSNs, accentuate the need for new protocols that consider the specific constraints related to these newly developing WSNs.

REFERENCES

[1] Akyildiz IF, Su W, Sankarasubramaniam Y, Cayirci E. Wireless sensor networks: a survey. Comput Netw 2002;38(4):393–422.
[2] Takagi H, Kleinrock L. Optimal transmission ranges for randomly distributed packet radio terminals. IEEE Trans Commun 1984;32(3):246–57.
[3] Li J, Gewali L, Selvaraj H, Muthukumar V. In: Hybrid greedy/face routing for ad-hoc sensor network. Euromicro symposium on digital system design, 2004. DSD 2004, August. IEEE; 2004. p. 574–8.
[4] Chávez E, Mitton N, Tejeda H. Routing in wireless networks with position trees. In: Ad-Hoc, Mobile, and Wireless Networks; 2007. p. 32–45.
[5] Heinzelman WR, Chandrakasan A, Balakrishnan H. In: Energy-efficient communication protocol for wireless microsensor networks. Proceedings of the 33rd annual Hawaii international conference on system sciences, 2000, January. IEEE; 2000. p. 10.
[6] Abbasi AA, Younis M. A survey on clustering algorithms for wireless sensor networks. Comput Commun 2007;30(14):2826–41.
[7] Jung W-S, Lim K-W, Ko Y-B, et al. Efficient clustering-based data aggregation techniques for wireless sensor networks. Wirel Netw 2011;17(5):1387–400.
[8] Patil NS, Patil PR. In: Data aggregation in wireless sensor network. IEEE international conference on computational intelligence and computing research; 2010.
[9] Karthikeyan V, Sur S, Narravula S, et al. Data aggregation techniques in sensor networks. Osu-cisrc-11/04-tr60, The Ohio State University; 2004.
[10] Ozdemir S, Xiao Y. Secure data aggregation in wireless sensor networks: a comprehensive overview. Comput Netw 2009;53(12):2022–37.
[11] Wang Y, Attebury G, Ramamurthy B. A survey of security issues in wireless sensor networks; 2006.
[12] Newsome J, Shi E, Song D, Perrig A. In: The sybil attack in sensor networks: analysis & defenses. Proceedings of the 3rd international symposium on Information processing in sensor networks. ACM; 2004, April. p. 259–68.

[13] Karlof C, Wagner D. Secure routing in wireless sensor networks: attacks and countermeasures. Ad Hoc Netw 2003;1(2):293–315.

[14] Wood AD, Stankovic JA. Denial of service in sensor networks. IEEE Comput 2002;35(10):54–62.

[15] Perrig A, Stankovic J, Wagner D. Security in wireless sensor networks. Commun ACM 2004;47(6):53–7.

[16] Ozagur Sanli H, Ozdemir S, Cam H. In: SRDA: secure reference-based data aggregation protocol for wireless sensor networks. Proceeding of the 60th IEEE Vehicular Technology Conference, VTC'04, vol. 7, September 26–29; 2004. p. 4650–4.

[17] Ozdemir S. In: Concealed data aggregation in heterogeneous sensor networks using privacy homomorphism. Proceedings of the ICPS'07: IEEE international conference on pervasive services, Istanbul, Turkey; 2007. p. 165–8.

[18] Taban G, Gligor VD. In: Privacy-preserving integrity-assured data aggregation in sensor networks. Proceeding of international symposium on secure computing, SecureCom, Vancouver, Canada, August 29–31; 2009. p. 168–75.

[19] Agrawal R, Kiernan J, Srikant R, Xu Y. In: Order preserving encryption for numeric data. Proceedings of the 2004 ACM SIGMOD international conference on management of data, Paris, France, June 13–18; 2004. p. 563–74.

[20] Chan H, Perrig A, Song D. In: Secure hierarchical in-network aggregation in sensor networks. Proceedings of the 13th ACM conference on computer and communications; 2006. p. 278–87.

[21] Huang S-I, Shieh S, Tygar JD. Secure encrypted-data aggregation for wireless sensor networks. Wirel Netw 2010;16(4):915–27.

[22] Sandhya MK, Murugan K. In: Secure framework for data centric heterogeneous wireless sensor networks. Proceeding of: recent trends in network security and applications—third international conference, CNSA 2010, Chennai, India, July 23–25; 2010.

[23] Chen CM, Lin YH, Lin YC, Sun HM. RCDA: recoverable concealed data aggregation for data integrity in wireless sensor networks. IEEE Trans Parallel Distrib Syst 2012;23(4):727–34.

[24] Boubiche S, Boubiche DE, Bilami A, Toral-Cruz H. An outline of data aggregation security in heterogeneous wireless sensor networks. Sensors (Basel, Switzerland) 2016;16(4):525.

[25] Coulouris G, Dollimore J, Kindberg T, Blair G. Distributed systems: concepts and design. 5th ed. Boston, MA: Addison-Wesley; 2012.

[26] Akkaya K, Younis M. A survey on routing protocols for wireless sensor networks. Ad Hoc Netw 2005;3:325–49.

[27] Al-Karaki JN, Kamal AE. Routing techniques in wireless sensor networks: a survey. IEEE Wirel Commun 2004;11(6):6–28.

[28] Caruso A, Chessa S, De S, Urpi A. In: GPS free coordinate assignment and routing in wireless sensor networks. INFOCOM 2005. 24th annual joint conference of the IEEE computer and communications societies, proceedings IEEE, vol. 1, March. IEEE; 2005. p. 150–60.

[29] Dolev S. Self-stabilization. Cambridge, MA: MIT Press; 2000.

[30] Dijkstra EW. Self-stabilization in spite of distributed control. In: Selected writings on computing: a personal perspective. New York: Springer; 1982. p. 41–6.

[31] Demirbas M, Arora A, Mittal V, Kulathumani V. A fault-local self-stabilizing clustering service for wireless ad hoc networks. IEEE Trans Parallel Distrib Syst 2006;17(9):912–22.

[32] Johnen C, Nguyen LH. Robust self-stabilizing weight-based clustering algorithm. Theor Comput Sci 2009;410(6):581–94.

[33] Miyaji A, Omote K. Self-healing wireless sensor networks. Concurr Comput Pract Exp 2015;27(10):2547–68.

SENSING IN HEALTH AND WELL-BEING APPLICATIONS

TANGIBLE USER INTERFACES FOR AMBIENT ASSISTED WORKING

4

Antonio Xohua-Chacón, Edgard Benítez-Guerrero, Carmen Mezura-Godoy

University of Veracruz, Xalapa, Mexico

4.1 INTRODUCTION

Ambient intelligence (AmI) is an emerging discipline that brings intelligence to our everyday environments and makes those environments sensitive to us. AmI is built upon advances in sensors and sensor networks, artificial intelligence, software engineering, and other disciplines of computing. Two related subfields have emerged recently, namely ambient assisted living (AAL) and ambient assisted working (AAW). The first is related to environments that can offer assistance to their inhabitants while they are engaged in their everyday activities, while the second is related to work environments.

In AmI environments, users are expected to interact with the system in natural ways. Tangible user interfaces (TUIs) are a type of interface that links together physical and digital worlds, offering opportunities for promoting new kinds of physical interaction by manipulating objects instead of using a keyboard or mouse. In recent years, this field of research has been established in a significant way; however, TUIs have some limitations that need to be reduced for better performance in *ambient assisted systems*.

The objective of this chapter is to introduce the integration of tangible interfaces within an assisted working environment and provide new types of interactions for the users to develop their work.

This chapter is organized as follows. First, a general introduction to ambient intelligence, ambient assisted living, and ambient assisted working is presented. Then, a description of tangible user interfaces is developed, highlighting their characteristics as well as their advantages and limitations. Next, the concept of intelligent interfaces is presented, and finally some changes that must be considered for tangible interfaces to be part of ambient assisted working are explained.

4.2 AMBIENT INTELLIGENCE
4.2.1 GENERAL CONCEPTS

Ambient intelligence is an area of computing that is focused on empowering people's capabilities, through digital elements in the environment that are sensitive, adaptable, and responsive. An AmI system has three main functions (see Fig. 4.1): first, sensing of the environment; second, use of this information to reason about the environmental status; and third, actuation to change the state of the environment [1].

Intelligent Data Sensing and Processing for Health and Well-being Applications. https://doi.org/10.1016/B978-0-12-812130-6.00004-4

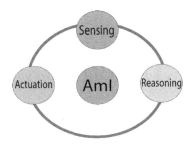

FIG. 4.1

Functions of an AmI system.

Detection of the environment is performed by sensors that monitor indicators such as temperature, lighting, noise, or vital signs such as heartbeats or brain activity. Reasoning occurs using the information obtained from such sensors; this phase can realize several types of reasoning such as

- *Modeling*: Models the behavior of the user; if it is possible to construct this model, it can be used to customize the behavior of the AmI software towards the user.
- *Recognition and activities prediction*: Attempts to identify the activities that the user is engaged in or to predict which ones will be engaged in, to anticipate user needs.
- *Decision making*: Artificial intelligence techniques are used to choose among several options available to address a certain activity or situation.
- *Spatial and temporal reasoning*: Focuses on implicit or explicit references to where and when significant events occur, for example, where the user is and for how long.

Acting is the way the system links to the real world, through intelligent devices or assistance, in some cases through the use of robots.

AmI has served as a basis for the development of two concepts: (a) ambient assisted living (AAL), which incorporates technologies that can be used to improve the health and well-being of people [2], and (b) ambient assisted working (AAW), in which the AmI concepts are applied to workspaces in offices [3] or production environments [4].

4.2.2 AMBIENT ASSISTED LIVING

Thanks to advances in medicine, life expectancy has been increased compared to previous generations. This has augmented the need for assistance for older people (see Fig. 4.2), due to loss of cognitive, auditory, visual abilities or age-related illnesses [2]. It is reasonable to think that elderly people are a vulnerable population sector and need more support than others due to these limitations; however, AAL is also useful for people without any disability to perform daily activities. The main objective of AAL [5] is to achieve benefits for

- *People*: Increasing their safety and well-being.
- *Economy*: Using limited resources with higher efficiency.
- *Society*: Offering better standards of living.

Because of the breadth of application coverage of the needs of older people, AAL environments are structured on three levels: hardware, middleware, and services (see Fig. 4.3).

FIG. 4.2

Needs for elderly people.

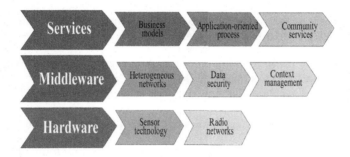

FIG. 4.3

Ambient assisted living environments levels.

At the hardware level, sensor technology is included; for example, wearable vital sensors are important for monitoring the state of health (it is usually a continuous monitoring) and recording. For a meaningful processing of this information, a continuous networking of sensors is a prerequisite, for example, circuit-based procedures such as building automation buses or local wireless networks to mobile communications. The diverse requirements arising from real scenarios can be met with heterogeneous network structures with different communication technologies.

Middleware focuses on the use of networked sensors and actuators; it must be able to flexibly provide a variety of sensors for different services to be deployed. The core of such an infrastructure is context management, which aggregates the collected sensor data and interprets them with the help of domain-specific application knowledge, thus providing context information at a higher abstraction level [6].

At the services level, the basic idea is to be able to configure more complex service offerings from granular but independent functional services. Service design requires a goal-oriented approach at two levels: (1) the conceptual level, to identify which service for a given purpose and under which specific conditions should be offered, and (2) the process level, focusing on how the specific service creation and offer process must be designed in order to achieve the previously defined goals in an effective and efficient way.

To go deeper into AAL applications, one may consult Rashidi and Mihailidis [2], where a study is carried out on the technologies, tools, and techniques used in different AAL projects.

4.2.3 **AMBIENT ASSISTED WORKING**

Another important field is work places, where it is possible to find a great diversity of workforce personnel, as well as different types of activities that each must carry out. Each worker has, on the one hand, his own abilities that differentiate him from others: for example, analysis skills, teamwork, personnel management, decision making, among others. On the other hand, each has special knowledge: for example, the domain of some application for editing text files, knowledge of a particular language, or legal knowledge. Due to the current work formats that break with the traditional borders of time and space, and because over time people develop a certain number and degree of skills related to work performance, they require support systems to increase their productivity and adapt to their needs, anticipating or adjusting effectively and efficiently to changes in their abilities [7]. This is the domain of AAW, which can focus on meeting the needs of all people, including people with disabilities and elderly people who are still working.

The degree of availability of computer and communications infrastructure offered in the workplace is higher than that found at home, including access control systems, process control, wireless devices, and cameras, among others, so that it is possible to create an advanced infrastructure for AAW systems. In particular, three levels are identified (as in AAL) [7]:

- *Network*: Determines the flexibility required for communication, including Wi-Fi networks, Bluetooth, and sensor networks, among others.
- *Agents*: A large variety of agents are integrated such as machines, robots, workflow control systems, or building automation systems.
- *Mobile personal interface*: Varies depending on the individual requirements and work tasks.

All systems are connected, allowing sensors and actuators to exchange information to create an intelligent and flexible environment, and interaction with the worker can be achieved by means of direct detection or by interaction with agents or by operation of a personal mobile interface.

4.3 **TANGIBLE INTERFACES**

With tangible user interfaces (TUIs), the user interacts with a digital system through the direct manipulation of physical objects, which are directly linked to a certain functionality within the system, so that manipulation affects its behavior. These objects are called tokens [8], and they can be varied in form, material, texture, color, size, weight, and built-in technology, since tokens are enriched with sensors or other devices that allow communicating with the system and with other tokens.

The application field for TUIs has been varied; they are used in applications such as entertainment [9], music [10], education [11], and health [12], among others. To have a more complete perspective on TUIs, one may consult Shaer [13], where a useful study including history, research trends, and background theories, as well as technologies, is presented.

4.3.1 **TAC PARADIGM**

A paradigm for describing and specifying TUIs was introduced in Ref. [8]. They mention that all manipulations that can be performed on tokens are subject to certain constraints, and that these limitations help the user to know how to manipulate the token and how to interpret the relationships that are created between them. This help given to the user must occur naturally, without forcing the user to think too much about the operation of the interface.

The TAC (token and constraints) paradigm captures the main elements of TUIs and addresses the conceptual challenges for building them. Terms used include

- *Pyfo*: A physical object that is part of the TUI. It can also be understood as a set of other pyfos: for example, a box can have six pyfos or sides.
- *Token*: A pyfo that can be manipulated with the hands and represents digital information or a computational function within an application. The user interacts with the token in order to access or manipulate the digital information.
- *Constraint*: A pyfo that limits the behavior of the token with which it is associated.

As mentioned previously, the interaction in tangible interfaces is provided by physical objects; however, the term *object* may be too ambiguous, since it has been used in several areas of computation, for example, in object-oriented programming. For this reason Ullmer and Ishii [14] propose to refer to objects as tokens. A physical object is considered to be a token only after it has been assigned to a variable within the system [15]. This means that a token represents something within the system and the manipulation of this affects in some way the digital information that it represents in the system. The physical properties of a token may reflect the nature of the information or function that it represents, but it can also provide the way in which it should be manipulated.

Constraints offer a reference frame for relative and spatial interpretation, besides the composition of a token with other pyfos associated with the same constraint. The physical properties of a constraint guide the user in understanding how to manipulate the token and how to interpret its behavior. A constraint limits the behavior of the token in three ways:

1. The properties of a constraint, such as orientation, material, texture, among others, suggests to the user how to manipulate and not manipulate the token associated with it.
2. The constraint limits the physical space interaction of the token. When the token is manipulated within the limits of a constraint, the interaction space is limited by the constraint.
3. The constraint serves as a frame of reference for the interpretation of the token and the composition of the constraint. The composition of a token and constraint must be interpreted in spatial terms, such as coordinates and numerical values, or in relative terms such as the positions first, left of, or next to, among others; in both cases, the compositions are interpreted with respect to a frame of reference.

A *variable* is digital information or a computational function within an application. Some variables are coupled to tokens while others are semantic variables in the application.

A TAC is the relationship between a token and its variable or one or more constraints; often this is a temporary relationship. A relationship is defined by the designer and an instance is created either by the designer or by the user. The physical manipulation of a TAC is the manipulation of a token with respect to its constraint and its computational implications.

The TAC model contains five main properties: coupling, relative definition, association, computational interpretation, and manipulation:

1. *Coupling*: A pyfo must be coupled with a variable so that it can be considered as a token; the designer defines what type of variable should be associated with a certain pyfo. The actual coupling of the variable with the token can be done in design time or at runtime, by the designer and the user respectively.

2. *Relative definition*: Each pyfo can be defined as a token, a constraint, or both.
3. *Association*: A new TAC is created when a token is physically associated; a new constraint can be associated with an existing TAC. The tokens and constraints have a recursive structure, which allows a TAC to be a token or a constraint for other TACs.
4. *Computational interpretation*: The manipulation of a token with respect to its constraint also has a computational interpretation and therefore there is a change of state in the application (e.g., changing the value of a variable that affects some function or functions in the system).
5. *Manipulation*: Each TAC can be manipulated discretely, continuously, or both. The physical manipulation of a token is provided by the physical properties of its constraints.

4.3.2 CLASSIFICATION

Three main types of TUIs are identified (see Fig. 4.4) [16]:

(A) *Interactive surface*: Often, tangible objects are placed on flat surfaces, and location and relationship between objects can be interpreted by the system. An example of this type of interface is URP [17].
(B) *Constructive assembly*: Modular elements that can be connected to similar ones to create models of physical constructions. Both spatial organization and action orders can be interpreted by the system, an example of which is Topobo [18].
(C) *Token + restriction*: Combines two types of physical-digital objects. Constraints provide a structure that limits the movement and position of tokens, and at the same time serving the user as a tactile guide. Constraints can express and reinforce a syntax for interactions. An example of this is the slot machine [19].

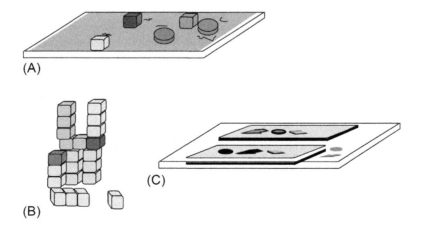

FIG. 4.4

Different types of tangible interfaces. (A) Interactive surface, (B) constructive assembly, and (C) token + restriction.

Based on Ullmer B, Ishii H, Jacob RJK. Token+constraint systems for tangible interaction with digital information. ACM Trans Comput Hum Interact 2005;12(1):81–118. doi:10.1145/1057237.1057242.

In spite of this classification, identification is not always easy, because sometimes the same tokens can serve as a restriction on others, or restrictions can be placed within others.

4.3.3 **ADVANTAGES**

The nature of TUIs represents a paradigm shift of interaction, accompanied by certain benefits that could be said to be inherited when integrating objects. An example, from a noncomputational point of view, is the use of objects in physical kindergartens and elementary schools to introduce young learners to abstract concepts such as quantity, numbers, or fractions. Following are some of the benefits of TUIs identified in Ref. [13]:

* *Collaboration*: One of the objectives of TUIs has been to promote dialogue between domain experts and stakeholders, as well as to support collaborative work [17] not limited to a single location, but it can also be done remotely [20].
* *Contextualization*: TUIs can inhabit the same world as the user, that is, the interaction with these interfaces is located in real space and therefore always located in specific places, not only residing on the screen. This is because of the very nature of TUIs, which implies that the meaning of the interaction of tangible devices can change depending on the context in which they are found, and, conversely, they can alter the meaning of the place. Some examples are summarized in Ref. [21].
* *Tangible thinking*: The physical body and the physical objects with which it interacts both play a very important role in the formation of an individual's understanding of the world. Babies develop their spatial cognitive skills through locomotor experience; in Ref. [22] the potential benefits of TUIs in learning are presented. TUIs take advantage of this connection between the body and knowledge, facilitating tangible thinking, thinking through bodily actions, physical manipulation, and tangible representations.
* *Gesture*: Gestures are typically considered as a means of communication; however, studies have shown that gestures play an important role in the alleviation of the cognitive load in both adults and children and in the conceptual planning of the speech production. TUIs allow users to take advantage of thought and communication through unrestricted gestures while interacting with the system. In Ref. [23] an API is presented that helps to create tangible interactions.
* *Tangible representation*: There are different ways in which TUIs use physical objects as representations of digital information. Some application domains such as architecture, urbanism and chemistry have inherent topological or geometrical representations that can be directly used in TUIs. And in the case of other domains, such as economics, biology [24], or music [25], where these forms are not available, they can be represented in a tangible way through metaphors.

4.3.4 **LIMITATIONS**

The limitations of TUIs are also inherited and are mainly associated with objects and their constraints. TUIs are unlike traditional interfaces that offer relative ease in realizing changes of the digital objects, for example, changing object size or duplication of the object. Ullmer and Ishii [14] mention that these interfaces require a careful balance between physical and graphical expression, to avoid physical clutter and to take advantage of contrasting strengths of different representational forms. Another limitation type is intelligence integration, which can reduce the number of necessary interactions to accomplish a specific task or also allow TUI to be adapted to the user.

The limitations of TUIs result in the following challenges [13]:

- *Scalability*: Applications for small problems or datasets often do not adapt to more complex problems that require many parameters and large data, because larger representations require a bigger workspace.
- *Versatility and malleability*: Digital objects can be easily created, replicated, modified and distributed while physical objects are rigid and static (in some cases); however, flexible material technology is already being developed [26] that changes form and that has the possibility of serving as input and output information.
- *User fatigue*: The effects of weight, size, and shape of tangible objects should not be underestimated; ergonomics should be considered as well as the variety of long-term interactions that are necessary to perform the tasks [27].
- *Automatic user identification*: Although it is true that devices can be adapted to perform some type of authentication (similar to those used by traditional interfaces, e.g., fingerprint reader), there is no way that the interface can identify the user automatically.
- *Customization and adaptation*: A desirable feature in the current interfaces is customization: an interface capable of identifying the user, and based on it, adapt to their preferences. This is not observed in tangible interfaces.
- *Task assistance*: The ability to assist the user in performing routine tasks, but without taking away the control.

4.4 INTELLIGENT INTERFACES

Throughout the history of *user interfaces* (IUs), different styles of interaction have been designed, such as command line (where the user writes the instructions that the computer should perform), navigation menus, and direct manipulation, this last being one of the most popular. The success of this type of interaction is due to the constant feedback provided to the user regarding the task being performed, which allows changes or corrections to be made quickly [28]. However, in this style of interaction the user performs all tasks and controls all events, which can cause a cognitive overload, as well as loss of usability.

Intelligent user interfaces (IUIs) are a subfield of what we know as *computer human interaction* (CHI). In Ref. [29] it is mentioned that intelligent interfaces have been proposed to solve the problem of cognitive overload as well as the saturation of information, and to provide aid on the use of complex systems. IUIs are also used to provide individualization and personalization of systems and, with this, increasing their flexibility.

Artificial intelligence (AI) is a field that influences IUIs, but also psychology, ergonomics, human factors, cognitive science, and social sciences are involved. In this chapter, only a few topics regarding AI are discussed.

4.4.1 GENERAL ARCHITECTURE

A general IUI architecture [30] is shown in Fig. 4.5. Input from the keyboard, mouse, microphone, camera, or possibly some other devices is recorded and then preprocessed (*processing* step includes event tagging and other important entry characteristics). After each input mode has been analyzed, the separate modes are merged and evaluated together. In some cases, it is desirable to merge the stream of inputs before processing them, depending on the application and the characteristics that need to be detected. Once it is known where the input comes from, the necessary course of action can be determined. First, it is necessary to evaluate what to do in the current situation. If information is missing or if the user has requested information (e.g., if the recorded voice contains a user question) this information will be requested from the application or from an external source. Usually an inference mechanism adopts conclusions and updates the system information: the user model, interaction history, and information about the application domain. Once all the necessary information is available and up-to-date, the system must decide the best alternative for action. This happens in the *interaction management* component, where usually some kind of adaptation of the interface is chosen. Often evaluation and adaptation occur at the same time using an inference engine for both, so the line between the processes of *evaluation* and *adaptation* is not very clear. The action chosen must still be generated, which occurs in the *output generation* step. Most IUIs can be created with this architecture, although it is often not necessary to explicitly model all parts.

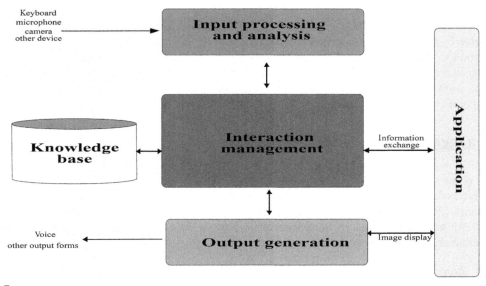

FIG. 4.5

General IUI architecture.

Based on Ehlert P. Intelligent user interfaces: introduction and survey. Research report DK S03-01/ICE01. Delft: Delft University of Technology; 2003. Available from: https://www.researchgate.net/publication/2566518_Intelligent_User_Interfaces_Introduction_and_Survey.

4.5 **TANGIBLE INTERFACES FOR AAW**

As mentioned previously, one of the main characteristics of AAW is that it can be adapted to the needs of people, based on detection of the user or interactions that they make with the agents (see Section 4.2.3). However, as mentioned in Section 4.3.4, tangible interfaces have certain limitations, such as adaptability, context sensitivity and task attendance, and cognitive load. On the other hand, intelligent interfaces (IUIs) try to solve these limitations by using artificial intelligence techniques (also present in AAL and AAW) that allow them to process multimodal inputs, offer help to the user, perform tasks automatically, adapt to the user, or generate responses in a multimodal way, among others.

Particularly, because the interaction type in TUIs is direct manipulation, this can increase the user cognitive load, in addition to other problems related to objects, such as the limitation of the workspace, number of tokens that the user can use to do a task, and the physical reproduction of these tokens.

Despite these limitations, tangible interfaces offer a different style of interaction than the traditional ones that have been used for several years (keyboard and mouse), and they can be useful for people with some type of disability as well as for those who do not have a disability. In addition, the feedback that the user can receive from a tangible interface is no longer only through a screen, but can also be in physical forms such as vibration or movement of the tokens.

There are some TUIs that could be used in work environments, to perform specific actions such as database queries [31]. These authors present two TUI prototypes that use tokens to represent database parameters; these tokens have a physical constraint associated with them. One prototype is Parameter wheels; this uses a small cylindrical token, each one containing RFID tags, and can represent both continuous and discrete values. Another alternative project is presented by Jofre et al. [32], in which they implement a tabletop with a 2D screen display, and the database queries are made by placing tokens on the tabletop by the user. In this case the constraint is the tabletop dimensions. For other work fields [17], they introduce URP, a TU system for urban planning, in which the physical architectural models are placed on an ordinary surface; the system allows emitting of precise shadows for arbitrary times of day, as well as wind flow. Physical telepresence [33] is another example in which a work area is shared and can be modified by a remote user.

4.5.1 **BACKGROUND**

Table 4.1 shows the elements of an IUI that were detected in different types of TUIs that were analyzed. It can be observed that all of them lack a *knowledge base*. This is due to the fact that the approach sought in each type of work is different. For example, they seek the tangible manipulation of the information visualization [38] or to provide a hardware and software infrastructure for the movement of objects on a surface [39]; however, the rest of the elements are present, so it would be important to analyze the implementation of this *knowledge base* and evaluate its impact.

4.5.1.1 Actuation

One of the main characteristics of tangible interfaces is that they offer the possibility of feedback to the user, either in an intangible way—for example, sound or image—as well as in physical form through tokens. Poupyrev et al. [21] they call this *actuation* and it is defined as "put into action, move"; in other words, actuation is the change of state or physical properties that an object can go through in an automatic way.

Table 4.1 Presence of Elements of General IUI Architecture [30]

Title	Input P&A	Interaction Management	Output Generation	Knowledge Base
Pressure sensor imaging for tabletop tangible interaction (PSITTI) [34]	✓	✓	✓	✕
Mementos [35]	✓	✓	✓	✕
inTUIt [36]	✓	✓	✓	✕
Mechanix [37]	✓	✓	✓	✕
InfoPhys [38]	✓	✓	✓	✕
Actuated workbench [39]	✓	✓	✓	✕
Protrude, flow [40]	✓	✓	✓	✕

The types of feedback of tangible type can occur through the alteration of some of the physical properties of the tokens; for example:

- *Changing the spatial position of objects or their parts*, for example, their position or orientation in the workspace.
- *Change the speed of movement of objects or their parts*, for example, the speed of rotation, linear speed of movement, or its direction.
- *Change the texture of the object or its parts* on the surface that can be perceived by touch.
- *Change the force applied to the user*, for example, the torque.
- *Change the color of the token.*
- *Change the weight.*
- *Change the volume.*
- *Change the shape.*

The preceding is derived from a limitation observed in most applications of tangible interfaces: physical objects are rigid and static, while on the other hand, digital objects are easily created, modified, or reproduced. Therefore, researchers and designers have taken the next step in the search to create interfaces where physical objects are not only linked to digital attributes and information, but also can be dynamic and alter their physical properties depending on the interface, user, or environment state.

Actuated Workbench [39] is an example of actuation interactive devices, which consist of several parts that can be reorganized by themselves in the workspace and which consist of a 2D array of electrical magnets, which manipulates the force and the shape of the magnetic field. This allows one or more objects on the surface to be moved and organized in any 2D pattern.

Table 4.2 presents some works that focus on the modification of some type of physical property of objects, through the use of different materials and electronic devices. The *OnObject* project allows novice end users to transform everyday objects into gestural interfaces, just by adapting a sensor, adding a behavior to the tagged object, and specifying the desired response; one of the facilities it offers is copying the behavior and the associated response from one object to another. *Haptic Edge Display* consists of an arrangement of actuated pins located around the screen of a cell phone, called tactile pixels or taxels. These taxels change their height individually, providing a haptic feedback to the user. *ZeroN* is a

Table 4.2 Works Focused on Change Physical Properties

Title	Color	Position	Weight	Form	Texture	Copy	Volume
OnObject [41]	X	X	X	X	X	✓	X
Haptic edge display [42]	X	X	X	✓	X	X	X
ZeroN [43]	X	✓	X	X	X	X	X
ChainFORM [44]	✓	✓	X	✓	X	X	X
Weight and volume changing device [45]	X	X	✓	X	X	X	✓
aeroMorph [46]	X	X	X	✓	✓	X	✓

tangible representation of the virtual world in 3D, using a magnetic control system to create a predefined 3D area where the object can be moved; it also uses a tracking system and a visualization system that projects images on the object that is levitating. *ChainFORM* is an active, modular, linear hardware system; the modules are connected to form a chain. They are formed by sensors that detect touch on several surfaces, detect angles, perform motor actuation, and offer visual feedback. One application of this work is the adequacy of these for the human body, even detecting an incorrect posture of the back. The *Weight and Volume Changing Device* by injecting or removing liquid metal with the aid of a bidirectional pump can modify the mass and volume of an object simultaneously. The *aeroMorph* device presents a mechanism that can create behaviors of change of form, on diverse materials like paper, plastic, and fabric, making use of pneumatic movement for the activation of the change forms.

4.5.2 AI TOPICS FOR TUIS

Next, some related AI topics that are considered relevant to be implemented on tangible interfaces are described.

4.5.2.1 Knowledge representation

According to Ehlert [30], the central idea behind learning in any computer is to deal with unforeseen situations and unknown circumstances; for example, instead of asking the user what task to do, the interface should be able to identify (through various sensors) the tasks that the user performs, build a representation of the world and based on this representation propose the automation of any task that is performed frequently. There are three types of knowledge that would be very meaningful for an IUI to obtain:

- *Unknown information*: Refers to information not known in advance, for example, the preferences of a specific user.
- *Dynamic information*: Refers to everything that changes over time: for example, you can have a user model and its environment in an initial way, but this information will become obsolete due to the changes of preferences and the environment.
- *Difficult to program knowledge*: Refers to information that is very difficult to program manually but can be learned with examples.

In the case of a computer, the knowledge can be stored in memory by filling it with certain symbols, where each symbol represents an object or idea; this is called knowledge representation. The knowledge representation language must be able to express all the necessary descriptions in an appropriate and effective way; in addition, the application domain determines the type of knowledge representation to be used. The reasoning system consists of data structures to store knowledge, as well as the necessary procedures to manipulate these structures and to be able to deduce new facts. The most commonly used types of representation in knowledge and reasoning systems are [30]

- predicate calculus
- semantic networks
- frames
- production systems
- Bayesian networks
- fuzzy systems

4.5.2.2 Models

Adaptation and problem solving are important issues that are addressed in research on artificial intelligence (AI), to achieve customization and flexibility.

In order to be flexible, many of the IUIs are based on adaptation or learning. Adaptation can occur based on the knowledge stored in models, which can be a

- Domain model, which contains data that the interface requires for the user to perform tasks.
- Task model, which stores all the tasks that the user can perform and the possible restrictions that can exist in those tasks.
- User model, which expresses the computer's need to understand humans and store this Information within their architecture [47]. This term represents a set of personal data associated with a specific user, usually containing personal information which is known as explicit information, because most of the time these data are provided by the user himself or herself; however, the most recent user models include user preferences, habits, and even more complicated data that are derived from stereotypes of users.

There are several categories of user models. However, there are four most-used models which are summarized in Ref. [47]: static models, nonstatic models, stereotypes, and adaptive.

It is important to consider the devices and type of tracking (individual or group) that a TUI needs in order to determine or construct user models.

4.5.2.3 Learning

The term *learning*, from a computational point of view, was defined by Meystel and Albus [47a] as "… a process based on the experience of intelligent system functioning (their sensory perception, world representation, behavior generation, value judgement, communication, etc.) which provides higher efficiency which is considered to be a subset of the (externally given) assignment for the intelligent system."

The types of learning are listed below:

- *Machine learning*: It is a subfield of artificial intelligence that seeks to improve performance in the task domain based on partial experience with a certain domain.

- *Reinforcement learning*: It is a method that tries to learn by exploring all possibilities within all available states (trial and error) and classifying the possible actions according to their opportunity and utility. Opportunities are determined by an evaluation mechanism that sends a signal of reinforcement to the control system. When the evaluation is performed by a human it is called supervised learning; on the contrary, when the evaluation is done automatically, it is called unsupervised learning.
- *Case-based learning*: Experiences are organized and stored as a case structure, then retrieved and adapted as necessary to the current situation. First, the current problem is classified, and then it uses the problem description to query similar cases in case memory, and then adapts the solution from an old case to the specifications of the new case. Next, this method applies the new solution and evaluates the results, and finally learns by storing the new case and its results.
- *Artificial neural networks*: Consist of nodes connected by links that have a certain weight; the learning occurs through the use of an error minimization procedure that adjusts the "synaptic" weights.
- *Genetic algorithms*: A population of individuals (solutions) is classified using a fitness function and the most suitable individuals are allowed to procreate; later individuals with low performance are extinguished with the generations until the population improves the quality of its set of solutions.

Learning occurs when stored knowledge is changed to reflect new data. Due to inherent difficulties in creating IUIs and the amount of knowledge engineering that is needed, most IUIs focus on a specific method of interaction or a well-defined application domain.

4.5.2.4 Personalization examples
An overview of web personalization is presented in Ref. [48]. Personalization is seen as the application of machine learning and data mining techniques for the development of user behavior models that can be used for predicting user needs and adapting future interactions in order to improve user satisfaction.

On the other hand, personalization in a television application is presented in Ref. [49], as allowing the user to customize the commercial cuts instead of using the predefined ones used by the television transmission streaming [50], and to use reinforced learning techniques in an information retrieval system through the use of keywords, in which the user specifies those of interest. Huang et al. [51] show an application that makes recommendations of chords to novice composers; here the recommendations are based on similar chords that are used in similar musical contexts. Paudyal et al. [52] present a system that translates gestures (sign language) in real time, allowing interaction with the computer; it can also be translated into a regular language for communication with another person.

4.5.3 INTELLIGENT TANGIBLE INTERFACES
The integration of AI techniques within a tangible interface would possibly allow a better integration of TUIs within an AAW system and function as one more agent.

The creation of a user model in a tangible interaction environment becomes a major challenge, due to the limitations mentioned in Section 4.3.4. A first approach would be to determine the level of experience of the user (using the interface) which is interacting with the system, based on the interactions that he makes with the different types of objects, the task that is being performed during the interaction and the time it takes to complete the task. With this information, it may be possible to create user stereotypes which could be novice, intermediate or advanced.

To achieve this, it is necessary to have a *domain model*, which will define the tasks that can be performed in the system and these tasks in turn define the interactions for each of them. Another important factor to consider is how the system will identify when the user is trying to solve a specific task or if the interaction with objects is due to a situation outside the system tasks.

Based on the user model, the next step would be to *adapt* the system to the user. For example, if the level of experience of the user corresponds to the novice level, it is likely that the user requires more help than an advanced level user. Then the tangible interface should be able to offer recommendations for the accomplishment of the tasks and at a certain moment the possibility of performing the task for the user. For example, if the system detects an interactions pattern with the objects and can identify these patterns as a subset of necessary interactions to solve a particular task, then the TUI should be able to deduce the missing interactions and offer the alternative to the user, i.e., whether that TUI should continue in automatic form with that task or not. Depending on the architecture of the interface, if the tokens have built-in technology that allows movement, the TUI should be able to move the tokens, simulating the movement by the user.

4.5.3.1 Possible future scenario

To illustrate the possibilities of TUIs in AAW, let us imagine an application in the field of architecture. In this application, an architecture firm has a network of TUIs distributed in its branches located in different cities. One of those TUIs located at the company's headquarters is used by an architect to build a model of a building using tokens that can change the shape, color, and size according to user needs. This TUI is able to register the states of tokens over time, which makes it possible to go from one state to a previous state of the work or even recreate the user actions. Imagine that it is immediately necessary to show the model in a different city. So, thanks to the existing network of the TUIs, the user can share the model and the remote TUIs can automatically replicate it by altering the physical properties of tokens by means of information about tokens states. In this way, the physical representation and changes of the model are almost instantaneous, without having to go through a 3D printing process, or the possible costs, damages, and time associated with shipping.

4.6 CONCLUSIONS

Thanks to business infrastructure, it is possible to build more complete AAWs that address the needs of people in their workplaces, offering support systems to increase productivity, with systems that are adapted according to the needs of each individual. However, despite their benefits, TUIs have some disadvantages, such as scalability of objects, malleability, and user fatigue, which can be both physical and cognitive. On the other hand, intelligent interfaces (IUIs), through the use of artificial intelligence techniques, try to solve some of the problems present in direct manipulation interfaces: for example, the creation of customized systems or execution of tasks on behalf of the user to lighten her/his cognitive load [53]. The integration of intelligence within a tangible interface could lessen the disadvantages described previously. However, such modification is not trivial, since it is necessary to consider several factors, such as the type of help that should be provided to the user and the means through which it can be realized, highlighting that it can be intangible or tangible through the tokens. If done in a tangible way, the properties mentioned previously must be considered in the tokens when designing them.

In the near future, the addition of TUIs to AAW can bring benefits such as (a) alternative user interactions that simplify communication with systems and new forms for user feedback, and (b) savings in resource utilization, given that traditional printers and paper or 3D printers to materialize digital information are no longer necessary, thanks to tokens that could alter their physical properties depending on user requirements. However, for this to happen it is necessary to overcome challenges such as: (a) enable malleability and change of texture of materials to enable the user to alter the physical form of the tokens or to choose their texture: e.g., to simulate the texture of glass, brick or metal; (b) introduce new forms of sensors, devices, or other elements to enable the movement of tokens and their communication with the system, with other tokens, or with other TUIs; and (c) incorporate information security mechanisms for user authentication to guard sensitive information as well as to protect token-token, token-TUI, TUI-TUI, and TUI-other systems communications.

REFERENCES

[1] Cook DJ, Augusto JC, Jakkula VR. Ambient intelligence: technologies, applications, and opportunities. Pers Mobile Comput; 2007. 277–98.

[2] Rashidi P, Mihailidis A. A survey on ambient-assisted living tools for older adults. IEEE J Biomed Health Inform 2013;17(3):579–90.

[3] López-Cózar R, Callejas Z. Multimodal dialogue for ambient intelligence and smart environments. In: Nakashima H, Aghajan H, Augusto JC, editors. Handbook of ambient intelligence and smart environments [December 2009]. Boston, MA: Springer; 2010. p. 559–79.

[4] Eschenbächer J, Wüst T, Thoben KD. Extended products supporting ambient assisted working approaches in production environments, In: Hoffmann P, Wewetzer D, Eschenbächer J, Herzog O, Thoben KD, editors. Proceedings of the Change 2009—ambient assisted working accessible and assistive ICT in enterprise environments (2010). Emden, Germany, 10–11 September; 2009. p. 54–74.

[5] Dohr A, et al. In: The internet of things for ambient assisted living. 2010 Seventh international conference on information technology: new generations, Las Vegas, NV; 2010. p. 804–9.

[6] Kunze C, et al. In: Kontextsensitive Technologien und intelligente Sensorik für Ambient-Assisted-Living-Anwendungen. Tagungsband zum 1 Deutscher Kongress Ambient Assisted Living AAL 2008; 2008. Available from, http://publications.andreas.schmidt.name/Kunze_Holtmann_Schmidt_Stork_AAL08.pdf.

[7] Bühler C. In: Stephanidis C, editor. Ambient intelligence in working environments. Universal access in human-computer interaction. Intelligent and ubiquitous interaction environments: 5th international conference, UAHCI 2009, held as part of HCI international 2009, San Diego, CA, July 19–24, 2009. Proceedings, part II, Berlin, Heidelberg: Springer Berlin Heidelberg; 2009. p. 143–9.

[8] Shaer O, et al. The TAC paradigm: specifying tangible user interfaces. Pers Ubiquit Comput 2004;8(5):359–69. https://doi.org/10.1007/s00779-004-0298-3.

[9] Tangible Media Group. In: Tangibles at play. ACM SIGGRAPH 2006 emerging technologies on—SIGGRAPH 06, Boston, MA; 2006. p. 32.

[10] Jordà S, et al. In: The reacTable: a tangible tabletop musical instrument and collaborative workbench. ACM SIGGRAPH 2006 sketches on—SIGGRAPH 06. Boston, MA; New York, NY: ACM Press; 2006. p. 91.

[11] Zuckerman O, Arida S, Resnick M. In: Extending tangible interfaces for education: digital montessori-inspired manipulatives. CHI05 proceedings of the SIGCHI conference on human factors in computing systems. Portland, OR; New York, NY: ACM Press; 2005. p. 859–68.

[12] Girouard A, et al. Tangibles for health workshop. In: Proceedings of the 2016 CHI conference extended abstracts on human factors in computing systems—CHI EA 16. New York, NY: ACM Press; 2016. p. 3461–8.

[13] Shaer O. Tangible user interfaces: past, present, and future directions. Found Trends Hum Comput Interact 2009;3(1–2):1–137.

[14] Ullmer B, Ishii H. Emerging frameworks for tangible user interfaces. IBM Syst J 2001;39(3.4):915–31. https://doi.org/10.1147/sj.393.0915.

[15] Shaer O, Jacob RJK. A specification paradigm for the design and implementation of tangible user interfaces. ACM Trans Comput Hum Interact 2009;16(4):1–39. https://doi.org/10.1145/1614390.1614395.

[16] Ullmer B, Ishii H, Jacob RJK. Token+constraint systems for tangible interaction with digital information. ACM Trans Comput Hum Interact 2005;12(1):81–118. https://doi.org/10.1145/1057237.1057242.

[17] Underkoffler J, Ishii H. In: Urp: a luminous-tangible workbench for urban planning and design. Proceedings of the SIGCHI conference on human factors in computing systems: the CHI is the limit. Pittsburgh, PA; New York, NY: ACM Press; 1999. p. 386–93.

[18] Raffle HS, Parkes AJ, Ishii H. In: Topobo: a constructive assembly system with kinetic memory. Proceedings of the SIGCHI conference on human factors in computing systems, Vienna, Austria; 2004. p. 647–54.

[19] Perlman R. Using computer technology to provide a creative learning environment for preschool children, Tech. rep., MIT AI Lab memo 360, Logo memo 24. 1976.

[20] Xiao X, Ishii H, Group TM. MirrorFugue: communicating hand gesture in remote piano collaboration. Funchal; New York, NY: ACM Press; 2011. 13–20.

[21] Poupyrev I, Nashida T, Okabe M. In: Actuation and tangible user interfaces: the Vaucanson duck, robots, and shape displays. Proceedings of the 1st international conference on tangible and embedded interaction TEI 2007, Baton Rouge, LA; 2007. p. 205–12.

[22] Marshall P. In: Do tangible interfaces enhance learning?. Proceedings of the 1st international conference on tangible and embedded interaction—TEI 07. New York, NY: ACM Press; 2007. p. 163.

[23] Nunes R, Rito F, Duarte C. In: TACTIC: an API for touch and tangible interaction. Proceedings of the ninth international conference on tangible, embedded, and embodied interaction—TEI 14. Stanford, CA; New York, NY: ACM Press; 2015. p. 125–32.

[24] Okerlund J, et al. In: SynFlo. Proceedings of the TEI 16: tenth international conference on tangible, embedded, and embodied interaction—TEI 16. New York, NY: ACM Press; 2016. p. 141–9.

[25] Peng H. In: Algo.Rhythm: computational thinking through tangible music device. Proceedings of the sixth international conference on tangible, embedded and embodied interaction. Kingston, ON; New York, NY: ACM Press; 2012. p. 401–2.

[26] Yao L, et al. In: PneUI: pneumatically actuated soft composite materials for shape changing interfaces. Proceedings of the 26th annual ACM symposium on user interface software and technology—UIST 13. St. Andrews; New York, NY: ACM Press; 2013. p. 13–22.

[27] Burgess-Limerick R, et al. Wrist posture during computer pointing device use. Clin Biomech 1999;14(4):280–6.

[28] López-Jaquero V, et al. In: Interfaces de usuario inteligentes: pasado, presente y futuro. VII congreso internacional de interacción persona-ordenador, interacción 2006; 2006.

[29] Höök K. Steps to take before intelligent user interfaces become real. Interact Comput 2000;12(4):409–26.

[30] Ehlert P. Intelligent user interfaces: introduction and survey, Research report DK S03-01/ICE01, Delft: Delft University of Technology; 2003. Available from, https://www.researchgate.net/publication/2566518_Intelligent_User_Interfaces_Introduction_and_Survey.

[31] Ullmer B, Ishii H, Jacob RJK. Tangible query interfaces: physically constrained tokens for manipulating database queries. Science 2003;3(c):279–86.

[32] Jofre A, et al. In: A tangible user interface for interactive data visualization. CASCON '15 proceedings of the 25th annual international conference on computer science and software engineering, Markham, Canada; 2015. p. 244–7.

[33] Leithinger D, et al. In: Physical telepresence: shape capture and display for embodied, computer-mediated remote collaboration. UIST '14 proceedings of the 27th annual ACM symposium on user interface software and technology. Honolulu, HI; New York, NY: ACM Press; 2014. p. 461–70.

[34] Holzmann C, Hader A. In: Towards tabletop interaction with everyday artifacts via pressure imaging. Proceedings of the fourth international conference on tangible, embedded, and embodied interaction—TEI 10. Cambridge, MA; New York, NY: ACM Press; 2010. p. 77–84.

[35] Esteves A, Oakley I. In: Mementos: a tangible interface supporting travel. Proceedings of the 6th nordic conference on human-computer interaction extending boundaries—NordiCHI 10; 2010. p. 643.

[36] Wiethoff A, Kowalski R, Butz A. In: inTUIt: simple identification on tangible user interfaces. Proceedings of the fifth international conference on tangible, embedded, and embodied interaction (TEI 11). Funchal; New York, NY: ACM Press; 2011. p. 201–4.

[37] Tseng T, Bryant C, Blikstein P. In: Mechanix: an interactive display for exploring engineering design through a tangible interface. Proceedings of the international conference on tangible, embedded, and embodied interaction (TEI 11). Funchal; New York, NY: ACM Press; 2011. p. 265–6.

[38] Frisson C, Dumas B. InfoPhys. In: Proceedings of the TEI 16: tenth international conference on tangible, embedded, and embodied interaction—TEI 16. New York, NY: ACM Press; 2016. p. 428–33.

[39] Pangaro GA. In: The actuated workbench: 2D actuation in tabletop tangible interfaces, interfaces. UIST '02 proceedings of the 15th annual ACM symposium on user interface software and technology, vol. 4(2). Paris; New York, NY: ACM Press; 2002. p. 181–90.

[40] Kodama S, Takeno M. In: Protrude, flow. Proceedings of SIGGRAPH'2001 electronic arts and animation catalogue. ACM; 2001. p. 138.

[41] Chung K. OnObject: programming of physical objects for gestural interaction. Master of Science in Media Arts and Sciences, Cambridge, MA: Massachusetts Institute of Technology; 2010.

[42] Jang S, et al. In: Haptic edge display for mobile tactile interaction. Proceedings of the 2016 CHI conference on human factors in computing systems. San Jose, CA; New York, NY: ACM Press; 2016. p. 3706–16.

[43] Lee J, Post R, Ishii H. In: ZeroN: mid-air tangible interaction enabled by computer controlled magnetic levitation. Proceedings of UIST 2011. Santa Barbara, CA; New York, NY: ACM Press; 2011. p. 327–66.

[44] Nakagaki K, et al. In: ChainFORM: a linear integrated modular hardware system for shape changing interfaces. Proceedings of the 29th annual ACM symposium on user interface software & technology. Tokyo; New York, NY: ACM Press; 2016. p. 87–96.

[45] Niiyama R, Yao L, Ishii H. In: Weight and volume changing device with liquid metal transfer. Proceedings of the 8th international conference on tangible, embedded and embodied interaction—TEI '14, Munich, Germany; 2014. p. 49–52.

[46] Ou J, et al. In: aeroMorph—heat-sealing inflatable shape-change materials for interaction design. Proceedings of the 29th annual symposium on user interface software and technology—UIST 16. Tokyo; New York, NY: ACM Press; 2016. p. 121–32.

[47] Alepis E, Virvou M. In: Object-oriented user interfaces for personalized mobile learning, vol. 64. Berlin, Heidelberg: Springer-Verlag; 2014. p. 25–9.

[47a] Meystel AM, Albus JS. Intelligent systems: architecture, design and control. New York: John Wiley & Sons, Inc.; 2002

[48] Anand S, Mobasher B. Intelligent techniques for web personalization. Intelligent techniques for web personalization, vol. 3. 169. 1–36.

[49] Chorianopoulos K, Lekakos G, Spinellis D. In: Intelligent user interfaces in the living room. Proceedings of the 8th international conference on intelligent user interfaces—IUI 03. New York, NY: ACM Press; 2003. p. 230.

[50] Glowacka D, et al. In: Directing exploratory search: reinforcement learning from user interactions with keywords. Proceedings of the 2013 international conference on intelligent user interfaces—IUI 13. Santa Monica, CA; New York, NY: ACM Press; 2013. p. 117–28.

[51] Huang CA, Duvenaud D, Gajos KZ. In: ChordRipple: recommending chords to help novice composers go beyond the ordinary. Proceedings of the 21st international conference on intelligent user interfaces. Sonoma, CA; New York, NY: ACM Press; 2016. p. 241–50.

[52] Paudyal P, Banerjee A, Gupta SKS. In: SCEPTRE: a pervasive, non-invasive, and programmable gesture recognition technology. Proceedings of the 21st international conference on intelligent user interfaces. Sonoma, CA; New York, NY: ACM Press; 2016. p. 282–93.

[53] Stephanidis C, Karagiannidis C, Koumpis A. In: Decision making in intelligent user interfaces. Proceedings of the 1997 ACM international conference on intelligent user interfaces (IUI97). Orlando, FL; New York, NY: ACM Press; 1997. p. 195–202.

AMBIENT ASSISTED WORKING APPLICATIONS

5

SENSOR APPLICATIONS FOR INTELLIGENT MONITORING IN WORKPLACE FOR WELL-BEING

Pablo Pancardo, Miguel Wister, Francisco Acosta, José Adán Hernández

Juarez Autonomous University of Tabasco, Cunduacan, Tabasco, Mexico

5.1 INTRODUCTION

The term ambient intelligence (AmI) was introduced by the Information Society Technology Advisory Group. The concept was created to designate environments where humans are surrounded by intelligent interfaces, supported by computer and network technology, living with quotidian objects such as clothing, furniture, vehicles, roads, and intelligent materials [1].

Scenarios with AmI should be aware of specific characteristics of human presence and its activities to deliver intelligent information services, both by the user's instructions and autonomously proactive, given the specific situation and context that arise [1]. An intelligent service is one that obtains information from the context using sensors, integrates it, and processes and produces useful knowledge for the user [2]. Hence a context-sensitive system requires being able to "recognize and measure" (usually through sensors), environmental values, user profile, and activities performed. The foregoing in order to offer value-added services should be supported with knowledge acquired from the context. AmI technological proposals should be discreet (interaction must be relaxed and enjoyable for the person) and should not imply an obstacle.

The AmI concept has been adopted and it is related to scenarios of everyday life, giving rise to what is known as ambient assisted living (AAL). AAL refers to intelligent care systems for a better, healthier, and safer life in preferred living conditions and encompasses concepts, products, and services that link new technologies and the social environment. The aim of AAL is to improve the quality of life, that is, the physical, mental, and social well-being of all people (with special attention to the elderly), at all stages of life [3].

AAL has received much attention due to the demographic growth of the adult population and the possibilities of care and support that it represents for the elderly. However, there is an equally important field of application in working life. In the labor context, people are asked for high mobility and flexibility, and people are required to work through advanced ages. But, although over time people develop more job knowledge, skills do not exhibit the same behavior. Therefore people need tailor-made support systems to maintain work efficiency and effectiveness, but with the necessary prevention or adjustment elements for changing skill levels [4].

Information and communication technologies can play a relevant role in contributing to the development of safe and healthy workplaces. In this sense, the AmI principles applied to the adaptation of workspaces have resulted in the emergence of a domain called ambient assisted working (AAW), which is an emerging area of AmI [4], where the proposed solutions are intended to address the specific needs of workers, taking into account values of environmental and physiological variables provided by sensors of various types.

Efforts realized by researchers in AAW are aimed at facilitating activities of the elderly or disabled within their working environment to make it a more healthy and comfortable environment. However, we believe that this area can be much broader and should include all people. The domain of AmI should not focus on the performance of people and should include studies on the welfare, safety, and health of workers in general. There are several examples of solutions in the literature that contribute to the well-being and health care of workers.

Migliaccio et al. [5] use acceleration and location sensors to identify incorrect postures when lifting a heavy object. That is, when using the correct position to lift an object from the ground, according to health specialists, the subject should flex the legs trying to keep the back straight, instead of bending to reach the objects on the ground. It is not advisable to flex the legs. Incorrect posture poses a risk of spinal cord injury. This study records this type of risky activity and the location where the activity is performed.

Senyurek et al. [6] state that they use accelerometers attached to the body to measure movement activities that might require alerting emergency services in the workplace. They mention that they monitor people working alone, which requires, in addition to detecting falls or excessive activity, also detecting other situations such as long periods of inactivity and unexpected positions of the worker. Activities such as running and jumping are monitored, as these activities are unusual in workplaces and could indicate an emergency situation. These are classified as excessive activities, for example, a long period of inactivity or an unexpected pose of the worker.

In the study of Koskimaki et al. [7], an inertial measuring device was placed on the wrist of a worker and the acceleration and angular velocity information was used to decide what activity the worker was performing in certain intervals of time. This activity information can be used for proactive training systems or to ensure that all necessary phases of the job are performed.

Other research work has been carried out related to AmI and particularly to AAW. These works are relevant because all workers are more or less exposed to health risks derived from their work activity, regardless of social, economic, or training status. Of course, some jobs are riskier than others. However, failing to take the necessary precautions to try to avoid the negative impact of accidents or illnesses produces overall losses for companies. Explicitly, the economic aspect of the organization is affected, since when a company does not have systems of monitoring and controlling occupational risks, this can lead to absenteeism, payment of high insurance premiums, loss of valuable personnel, payment for damages, etc. [8].

There are situations in which a system needs to be provided with particular data of the user's profile and activities performed in real time for offering effective solutions. The adjustment of a system with AmI, both to the context and to the users, is termed personalization [9]. Personalization is a characteristic that AmI systems must fulfill so that services offered are adapted to the needs of each user and the results offered are beneficial for those using the system, since these services must be created explicitly for each user and considering the current conditions of the context.

In particular, personalization in the area of health has led to the evolution of traditional preventive services in hospitals moving toward ambient intelligent systems that allow these services to be personalized and carried out in situ (home), at the proper time and in the proper way [10]. Another fundamental aspect of personalization in the area of health is that human beings are unique and therefore particular. What applies to one person could be completely different for another apparently equal in physiology and characteristics [11].

Consequently, systems are needed that incorporate customization mechanisms. Personalization is important because users are different, and beyond the adaptability of the interfaces it is necessary to have particular results. As the environment changes, the desired responses may change. Using the personalization process, the quality of the answers improves and the service life of the system increases.

Today important related advances to personalization are taking place, although most are directed toward educational scenarios, so the central focus is on user interfaces and presenting content [12], or toward models and specific users [13], for example, for supporting elderly or people with disabilities [10]. We think personalization in a broader sense is necessary, considering context awareness. A reference frame is needed allows visualizing systems that can have a great variety of desired answers.

In AAW systems, schemas are needed for customizing systems based on context. An architectural model serves as a reference for visualizing systems in intelligent environments. It must have a reference architecture and methods that allow implementing an AAW system, as well as the mechanisms for controlling (alerts, predictions, knowledge, etc.). Few research papers are focused on personalization in AmI systems for work environments and to our knowledge none with the approach as applied in this work.

An architecture and reference method helps to visualize advantages that AAW can offer beyond being limited to elderly or disabled people. A wide range of applications can be developed from architecture and reference methods: for example, surveillance of correct postures, surveillance of dangerous behaviors, and applications for various work areas (chemistry, mining, building, agriculture, trades in general). In particular, monitoring to ensure comfort and the consequent health improvements brings benefits to society as a whole, since it offers a proposal for compliance with occupational safety and health regulations.

A case study related to real-time thermal stress monitoring shows the importance of personalization to health. Thermal stress occurs when a person suffers adverse health effects due to being in an environment that is either very cold or very hot. When it occurs in an environment with high temperatures, it is called heat stress, and when it occurs in a workplace, it is labeled "heat stress in work environments." and when it occurs during a working day is particularly Heat Stress in Work Environments.

The implications of climate change lead us to expect devastating consequences for the population of tropical areas, such as occurred in India in 2015, when about 1700 people died due to heat related to climate change [14]. Studies in tropical areas demonstrate the importance of continuous monitoring of thermal stress, given the results that have been obtained. It is important to point out that, given the problems raised here, the motivation of research is toward health and wellness care using cutting-edge technology, rather than an economic purpose such as increased productivity or automation.

5.2 **STATE OF THE ART**

The foundation of the AmI vision is the fact that technological advances now allow the integration of electronics in the environment, thus supporting actors (people and objects) interacting with their environment in a natural and reliable way. In addition, there is a growing desire that the role of information and communication technology should not be limited to increasing productivity, but should also support people in their lives in terms of care, well-being, education, and creativity. In other words, new technologies should not increase functional complexity but should contribute to the experience of interacting with simple and easy-to-use products and services [15].

In Acampora et al. [9] and Aarts et al. [15], an AmI system is identified by several characteristics:

1. *Contextually aware*: It exploits contextual and situational information. Sensors embedded in the environment can determine the context in which certain activities take place, where the context provides meaningful information about people and the environment, for example, identity and spatial position.
2. *Personalized*: It is specific and meets the needs of each individual. The environment can recognize the users and adjust to individual needs to maximize the support. The automatic process to obtain profiles can capture the individual profiles of users through which it can set personalization and information filtering settings.
3. *Capable of anticipating*: It can anticipate the needs of an individual without the conscious mediation of the subject. While making the environment work without the conscious mediation of the user, it can also extrapolate behavioral characteristics and generate proactive responses.
4. *Adaptive*: The system adapts to the changing needs of individuals. The environment can change in response to the needs of the users. It can learn from recurring situations and changing needs, and adjust accordingly.
5. *Ubiquitous*: It is embedded and integrated into our everyday environments. Anywhere, anytime, any object.
6. *Transparent*: It remains discreetly in our daily lives.

The concept of AmI provides a vision of an Information and Knowledge Society, which emphasizes user-friendly systems, more efficient services, user empowerment, and support for human interactions. In this vision, people are surrounded by intelligent, intuitive interfaces that are embedded in all kinds of objects and in an environment that is able to recognize and respond to the presence of different individuals in a transparent, discreet, and often invisible way [16].

For development and implementation of AmI applications, microelectromechanical systems (MEMS) are essential. MEMS are mechanical and electromechanical devices and structures developed using micro-manufacturing techniques [17]. Technological advances in MEMS are very important and together with information and communication technologies, they enable the design of high-performance solutions, with low energy consumption and low economic cost, known as sensors. Sensors are small elements that can be embedded in everyday mobile objects and are able to communicate wirelessly. One of the main tasks of these network-integrated nodes is to gather context information (e.g., information on the environmental situation or the condition of an artifact) [18]. The main types of sensors used and the parameters they measure in ubiquitous and pervasive computing applications are motion, acceleration, light, proximity, audio, mechanical force, pressure, position, voltage, flux, radiation, temperature, and humidity [18, 19].

The domain of health and well-being has a high potential for contributing to market growth and the development of new sensor technology. Portable, wearable, and disposable sensors are used in health care, but also are used to measure physical activity and well-being. In many cases, they are used to monitor heart rate, blood pressure, respiration, and for diagnosis of specific diseases [19].

There are also emerging applications associated with environmental sensors, such as monitoring environmental conditions and structures. Other major smart cities' applications include active traffic management and interactive transport systems, intelligent networks for lighting and power supply, pollution monitoring, and weather forecasting. Most applications are enabled by wireless environmental sensor networks and cloud computing. Some types of sensors used are following.

- *Physiological sensors*. Sensor technologies for healthcare range from physiological monitoring (such as heart rate and blood analysis) to fall detection care, and rehabilitation applications. Sensors are used both in the home and in outdoor spaces, and some uses include telemonitoring and mobile health (mHealth). Some physiological sensors are used for the monitoring of sweating, body temperature, and heart rate.
- *Environmental sensors*. Intelligent systems gather a number of leading technologies and solutions to improve safety (at work, in buildings, in vehicles, etc.) and control of environmental conditions. Environmental sensors play a key role in safety and environmental detection applications [20]. Some of the best-known environmental measurement sensors are ambient temperature, relative humidity, solar radiation, wind speed, altitude, and latitude (GPS).
- *Activity sensors*. Human activities recognition is an important task for systems with AmI. Identifying a person's status is valuable information that can be used to feed into other systems. For example, the results of a fall detection system could be used to improve the recommendations of another system. Among devices for activities recognition, bracelets containing sensors have acquired great popularity because they are noninvasive technology and they make it possible to identify complex activities made up of several simple activities developed over time and space [21]. Some sensors for detection and recognition of activities include movement, acceleration, vision, cameras, and infrared [22].

All activities carried out in work environments, regardless of demographic, social, technological, and economic conditions, lead to occupational health and safety risks. Because of this, the analysis and evaluation of occupational risks play important roles in ensuring early identification and efficient prevention, thus contributing to the health and well-being of workers and strengthening the economic competitiveness of enterprises [23].

The International Labor Organization (ILO) is a tripartite United Nations agency with representatives of government, employers, and workers [8]. This tripartite structure makes the ILO a unique forum in which governments and social partners in the economies of its 185 member states can debate freely and openly to develop labor standards and policies.

The ILO Constitution establishes the principle that workers must be protected against diseases and accidents at work. However, for millions of workers the reality is very different. According to ILO statistics, 2.3 million people die each year from work-related accidents and diseases. An estimated 160 million people suffer from work-related illnesses and there are an estimated 317 million accidents per year related to work. The damage to workers and their families due to illness and accidents is incalculable [24].

In economic terms, the ILO has estimated that 4% of annual global GDP is lost as a result of occupational diseases and accidents. Employers face early retirements, loss of qualified staff, absenteeism, and high insurance premiums due to work-related illnesses and accidents. However, many of these tragedies can be prevented by applying good prevention, reporting, and inspection practices.

In order to improve the quality of life in work scenarios, a series of steps should be taken to ensure that workers are protected against illness or damage to health due to their work activities. The ILO recommends the introduction of a culture of preventive health and safety and the promotion and development of relevant instruments and assistance techniques [23].

In this sense, the emergence of the technological discipline of AAW becomes relevant, since it deals precisely with the implementation of the AmI vision in work environments, designing systems supported by sensors and computational techniques to offer services that facilitate the access, movement, and operation of all workers; provide support services for their activities; monitor and control environmental conditions; and seek safety and health through the prevention of damage and disease [3].

Therefore, the European Alliance for Innovation in AAL [3] mentions a number of work-related aspects that need to be covered for activities to be carried out in a wellness-supporting and healthy environment. These are

- access to workplaces,
- ensuring environmental conditions at work,
- support for work,
- prevention of diseases and damage to health, and
- safety and health regulations.

Based on previous research, in which AAW aims to support the productivity of older people, we define a new dimension of AAW, which we conceive as a work environment with the necessary technology of measurement, processing, and delivery of results that enables workers to carry out their activities based on intelligent automation of the environment toward productivity, collaboration, welfare, safety, and health, with transparent, discreet, and user-friendly interactions between systems and users.

Due to their advances, intelligent systems have become an integral part of many people's lives and the services they offer make them capable and varied. However, users expect to be able to customize a service to meet their needs and they do not accept a generic solution. So, intelligent systems must be able to customize the services offered to users, depending on their specific profiles [25].

Many researchers have established that the use of AmI in health care is very important, as it allows development of solutions based on individual medical conditions, such as physical or mental disabilities, chronic diseases, or rehabilitation situations, leading to personalized care. Developments in infrastructure and technology to achieve healthcare purposes include intelligent environments, assistance robots, portable sensors, and smart clothing [9]. Personalization and health care in work environments are necessary because of contemporary situations such as stress, workload, and occupational hazards [26].

5.3 AAW APPLICATIONS

This section describes the main aspects of two applications. The first one, "Heat Stress and Alarms," was partially reported with detailed explanation of the method and implementation in [27]. In the second application, "Classification of Perceived Exertion in Workplace Using Fuzzy Logic," we argue the advantages of using fuzzy logic to extend the reported proposal in [28].

The proposal in [27] was focused on the design and development of real-time personalized monitoring for more effective and objective estimation of occupational heat stress (OHS), taking into account the individual user profile, fusing data from environmental and unobtrusive body sensors. Formulas employed in this work were taken from different domains and were joined in the proposed method. It is based on calculations that enable continuous surveillance of physical activity performance in a comfortable and healthy manner. We found that OHS can be estimated by satisfying the following criteria: objective, personalized, in situ, in real time, just in time, and unobtrusive. This enables timely notice to workers for making decisions based on objective information to control OHS.

The work in [28] highlights the importance of using custom methods to estimate the physical exertion when people are doing physical work. The proposed application is based on data from heart rate sensors and work experience in performing a given activity. Reported results show how important it is to use personalized measurements for more objective effort estimation.

5.3.1 HEAT STRESS AND ALARMS

From workload and $WBGT_A$, the system can show the user, once per minute, if he/she is under OHS. To determine the level of OHS, we took the rating label that implies greater intensity of effort between movement (three levels of workload) and relative cardiac cost (RCC). The scale for RCC has also been grouped into three levels accordingly to ISO. Then, based on the ISO table (Table 5.1), the person is advised of his/her stress condition. Table 5.1 (ISO 7243) is used as a reference to determine when a worker has heat stress.

Personal estimation of workload levels is going to be the difference between visual appreciation and personalized heat stress estimation. For example, two workers in the same environmental conditions doing the same physical activities may differ in workload level estimation, leading to different results in OHS estimation.

Apart from measuring heat stress level (HSL), the devised solution also includes reactivity features, since it alerts the user when HSL is surpassed. It also notifies the worker's supervisor if the worker ignores warning messages and the recommended HSL is exceeded for a period.

Fig. 5.1 shows different GUI screens for the system prototype. Fig. 5.1A shows the personal data capture screen needed in the OHS personal calculation. This screen is intended to appear on the

Table 5.1 Wet Bulb Globe Thermometer (WBGT) Limit Temperature (°C) Values in Accordance With Workload

Work-Rest Regime	Workload		
	Light	**Moderate**	**Vigorous**
Continuous work	30.0	26.7	25.0
75% work + 25% rest; each hour	30.6	28.0	25.9
50% work + 50% rest; each hour	31.4	29.4	27.9
25% work + 75% rest; each hour	32.2	31.1	30.0

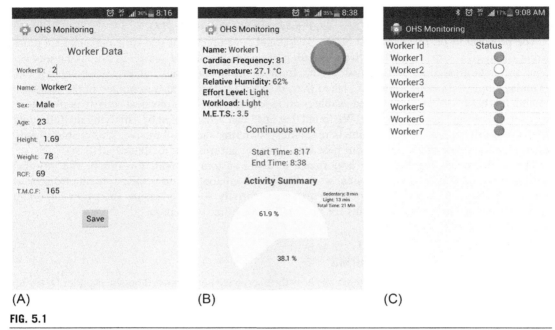

FIG. 5.1

(A) Worker data capture; (B) worker activity information; and (C) supervisor checklist.

supervisor's smartphone. In Fig. 5.1B, we can see the GUI screen with the user and ambient information, the calculated user effort level, the workload, and metabolic equivalents (METS), as well as a user message recommendation regarding the user's activity status (the user can continue working in this case) and a summary of activity data. The screen in Fig. 5.1C is intended for the supervisor, offering a general view of the worker's OHS status, helping him to make decisions for special attention when needed.

Fig. 5.2 depicts different OHS values corresponding to monitoring in the morning. The green light means that a user is in his/her comfort zone; a yellow light corresponds to a user being in the threshold of his/her comfort zone, and the red light shows a user suffering from OHS. The recommendation messages for each user correspond to the ISO values for workload shown in Table 5.1.

Figs. 5.2 and 5.3 show the impact of ambient temperature and relative humidity in the OHS calculation, because of the ISO compliance.

Fig. 5.3 shows different OHS statuses corresponding to monitoring in the afternoon.

The objective of this project was to achieve monitoring for an individualized OHS estimation. We were not trying to reach generalized proposals based on average values of previous studies and profiles of people type, as proposed in the standards, but the input values for the method and the resulting values are for one specific person working in situ in normal climate conditions.

This solution informs the worker and the supervisor, not necessarily in the same location, about the OHS status, taking values of temperature and relative humidity from the environment. The solution has distributed sensors, which are placed on each worker; the solution in ubiquitous, because it is present in

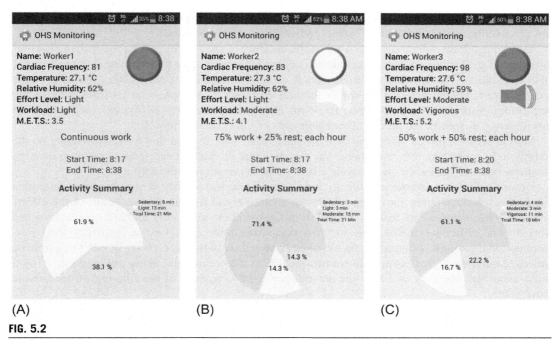

FIG. 5.2

Different OHS levels during morning monitoring: (A) without OHS; (B) warning alarm prior to OHS; and (C) alarm for OHS.

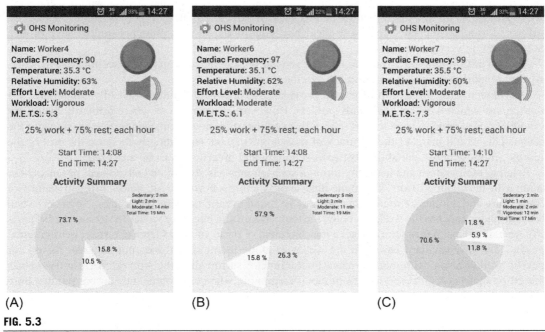

FIG. 5.3

Estimated OHS levels in the afternoon: (A) OHS for Worker 4; (B) OHS for Worker 6; and (C) OHS for Worker 7.

each smartphone and it communicates with the supervisor's smartphone to report workers' activities, and the solution knows the workers' status and environmental conditions.

Effort magnitude estimation was based on the intensity of the activity, not on recognizing the performed activity; further study may include activity recognition in order to determine its impact on the OHS calculation. The intensity of movements is calculated from the speed changes recorded by the accelerometer. This means that the use of the accelerometer is appropriate to estimate the effort in performing work activities. However, for work in which the activity is rhythmic and consistent (few changes in speed), the precision measurements of effort can be limited.

Motion sensors, placed on the hip or right-hand wrist, were used in this work. Depending on the activity to be measured, the smartphone was placed where the highest acceleration magnitudes were obtained, because obviously they provide a better characterization of the activities.

However, for better characterization in some types of physical activity, it is necessary to determine the data fusion techniques most appropriate for a precise and formal representation of the efforts involved. This includes data fusion obtained from sensors of the same kind, and it is mandatory when working with different types of sensors (movement, temperature, light, cardiac frequency, etc.). Furthermore, in this study, the smartphone has to be fixed tightly to the user's body in order to eliminate undesired movements, for example, movements which would occur if placed in a pocket. Proper fusion could avoid the necessity of fixing the smartphone tightly, because of the filtering of undesirable movements.

The low-cost approach of the proposed solution does not justify the cost of equipment for indirect calorimetry; instead, the values of METS were obtained from a linear regression equation, which was obtained from the calibration of the Samsung S4 smartphone with respect to the GeneActiv accelerometer (which is scientifically accepted). Therefore, the METS value is validated.

The limitation of the ISO 7243 standard to address a study in real conditions in the tropics, with temperatures above 31°C, even in conditions of light work, means that results would reach high OHS values that do not correspond to reality (which affects productivity). New studies should be performed to adapt the ISO table to tropical heat conditions. In fact, previous studies have proven ISO 7243 to be very cautious.

Another ISO limitation is how to capture special considerations (sickness, acclimation to work, physical condition, pregnant workers, etc.) in the method.

One objective of this proposal was to provide an economical solution basically supported by a smartphone, which workers commonly possess. However, in occupations where primarily handwork is done, it is desirable to have a wrist device that can measure movements and wirelessly transmit the results. This would increase the accuracy of measurements, but substantially increase the cost of the solution. In these cases, suitability between economy and precision must be analyzed.

Reading temperature and humidity through smartphones has at least one advantage, because these values can be obtained from the immediate area where each worker is, since different values of temperature and humidity (microclimates) may be found even in the same work area. However, when the smartphone touches the worker's body, the temperature and humidity values may be altered. This is why we claim that a better solution is the measurement of these variables from external environmental sensors. The ISO standard based on the wet bulb globe thermometer does not consider its use for specific worker areas, with many sensing microclimates surrounding the worker. We can use values offered by meteorological services from cloud computing when environmental sensors (temperature and humidity) are not available.

Based on the experimental results, we observed that there is no direct relationship between movements and cardiac frequency. For example, when we measured sweeping and glass window cleaning, movement magnitudes rose in a similar way; however, the cardiac frequency does not increase in the same proportion. On the other hand, when chairs are stacked, there are low movement magnitudes, less than sweeping and cleaning, but unlike these activities, the cardiac frequency increases significantly. This justifies the appropriateness of cardiac frequency as an indicator of physical exertion.

The results obtained from the smartphone are reliable and competitive compared with those reported in the literature. Moreover, it is a low-cost, in situ, noninvasive solution that can be deployed in real working environments.

Indeed, the results obtained from the sensors reflect the intensity and frequency of workers' physical activities. We verified that user relative cardiac cost is a good indicator of effort level. The cardiac frequency monitor is suitable for activities with less movement but that demand effort involving oxygen consumption.

Security and data privacy are sensitive issues in information systems within the AmI domain [29], due to the information exchange between components. In addition to this, when it comes to personal data relating to health and job performance, the situation may be even more critical. In our proposal, we request written consent from the workers to handle their data and assure compliance with local laws on data protection.

In our opinion, the proposed monitoring represents a convenient balance between social, economic, and productive interests of enterprises to ensure work quality and the welfare of their workers. Personalized monitoring for estimating heat stress is more effective and less invasive than ISO methods because it uses unobtrusive sensor technology to capture in real-time environmental parameters and effort intensity.

We found that our monitoring proposal can offer to users a clear vision to interpret energy expenditure and the arduousness index in order to estimate OHS and employ preventions. Some other authors have validated the advantages of sensor technology for activity recognition and to estimate energy expenditure in workplaces [30, 31].

Our proposal has important advantages. For instance, it is easy to implement, low cost, and employs personalized, real-time measurements instead of table values created from a standard person profile. Furthermore, it is universal and is deployable ubiquitously; its basic infrastructure is a mobile phone and it enables self-care and timely alerting of workers for making decisions about OHS control based on objective information. Finally, but no less importantly, it can be used in situ for continuously monitoring the workers. Wearing a smartphone on the hip is suitable for monitoring work activities that involve a balanced use of the body and limbs. For activities with hands, we must use a sensor bracelet to increase precision.

Our solution is a good example of how off-the-shelf mobile, personal health and sport monitoring devices, sensibly combined, can prove useful in a new generation of nonintrusive, effective mechanisms to monitor health variables, in this case, heat stress level. It also shows how context-awareness, personal sensing, and mobile computing can enable AAW.

Further work will consider recent wristbands containing sensors, such as skin temperature and sweating rate, that would offer new opportunities. It would be interesting to log data across time to analyze correlations with future ailments or diseases, and; even to build a health monitoring prevention system. New technologies and solutions should also be developed to carry out preventive actions.

5.3.2 CLASSIFICATION OF PERCEIVED EXERTION IN WORKPLACE USING FUZZY LOGIC

In this AAW application, we used fuzzy logic to obtain the effort perceived by workers when they perform physical activities of their work, since this facilitates the decision making of supervisors regarding worker allocation in the appropriate job. Commonly, ergonomic methods are generic and they do not consider the diffuse nature of the ranges that classify the efforts. So, a personalized monitoring that allows a real and efficient classification of the perceived individual is important.

Perceived effort is a well-developed mechanism that humans possess for sensing the strain involved in their physical effort. It is the act of detecting and interpreting the sensations arising from the body during physical exertion [32]. Continuous measurement of physiological parameters in individuals while performing daily or labor activities allows health and well-being preservation or improvement. The estimation of worker physical efforts in workplaces can be used for allocation of an employee in the appropriate position, adequacy of physical activities inherent to a type of work, prevention of accidents due to work demands, diseases prevention related to physical demands, etc.

In this sense, the habituation of a worker to perform a specific activity is needed, because we must consider if this worker has the skills needed to perform the activity well; that is, good performance in the execution of an activity may depend more on habituation than on other factors such as good physical condition. The allocation of work posts can be conditioned by the performance of the worker, therefore of their habituation.

On the other hand, cardiac cost and metabolic expenditure (elements for estimating effort) have traditionally been calculated using formulas and generic tables [33]. There are activities that can compromise the well-being and health of workers, for example, activities that require great physical effort, so the use of a personalized method of monitoring physical activities becomes more important.

An important tool in decision making is fuzzy logic and in accordance with [34], there are two important reasons to employ fuzzy logic: (1) data obtained from sensors' measurements could be imprecise and imperfect, (2) fuzzy logic can deal with imprecision and uncertainty due to its properties of performance and intelligibility necessary for the classification process.

A typical fuzzy logic inference system has four components: the fuzzification, the knowledge base (rules and fuzzy sets), the inference engine, and the defuzzification [35]. Fig. 5.4 shows those main fuzzy inference system steps.

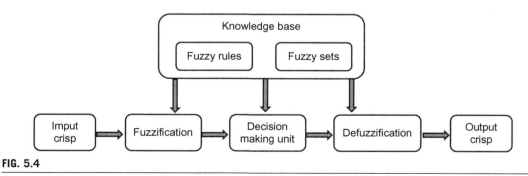

FIG. 5.4

Fuzzy inference system steps.

5.3.2.1 Habituation

Habituation is a learning form in which an organism decreases or ceases its responses to a stimulus after repeated presentations [36]. In a perceived exertion context, habituation is about how much a person has repeated a physical activity. Habituation as a state of training affects heart rate [37]. In a labor context, it refers to how frequently workers perform a specific physical activity related to their job.

Habituation to the performance of physical work activities is important because a person not accustomed to performing a specific physical activity has a perceived effort of about 20% higher than a person accustomed to performing such an activity [38].

5.3.2.2 Heart rate-based methods to estimate physical effort

Chamoux [39] proposes a lesser-known method that, as far as we know, is not frequently used. This method requires measuring resting and maximum heart rate for each person, taking into account several physiological parameters.

Effort levels for a worker according to Chamoux's method are shown in Table 5.2.

The calculation of effort by combining heart rate and fuzzy logic offers a more appropriate estimation of the effort since it allows inclusion of variables such as habituation that modify the perception of the effort of a worker in a personalized way. This can be offered as a tool at work to improve worker performance. Pancardo et al. [28] propose the method "Fuzzy Personalized Chamoux-based Method (FPC)" in which they use this combination.

FPC uses an inference engine fed by defined rules, where the level of habituation and the degree of membership to relative cardiac costs (RCC) are part of the rules, as shown in Fig. 5.5.

The last step is the defuzzification phase, where the perceived exertion is obtained.

5.3.2.3 Experiment

To observe the method's performance, the following experiment was undertaken. The tests were conducted on a university campus, and the subjects were potential janitors and janitors. In the experiment, a group of 20 research participants conducted a series of work activities and heart rate measurements were taken during those activities. These datasets were collected using a population of 20 participants: 11 males (28.4 ± 8.5 years, 171.8 ± 4.7 cm, 77.6 ± 13.09 kg, BMI 26.26 ± 3.77) and 9 females (28.7 ± 5.97 years, 161.1 ± 3.5 cm, 65.1 ± 11.4 kg, BMI 25.06 ± 4.45).

Table 5.2 Different Levels of Effort for RCC Under Chamoux

RCC	RCC Level	Effort
0–9	RCC1	Very light
10–19	RCC2	Light
20–29	RCC3	Slightly moderate
30–39	RCC4	Moderate
40–49	RCC5	Slightly heavy
50–59	RCC6	Heavy
60–69	RCC7	Intense

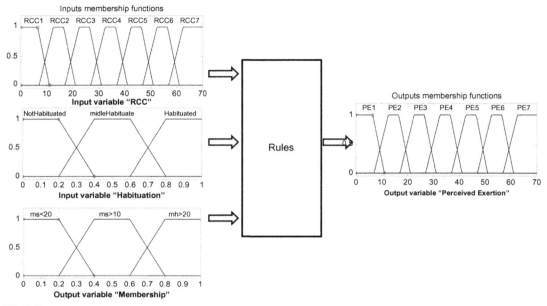

FIG. 5.5

Fuzzy system.

5.3.2.4 Results

In order to compare the resulting values of all methods tested, we made a mapping of Borg's perceived exertion values with labels used in the Chamoux method (Table 5.2). Scales 6–7 are No Exertion (NE), 8–9 are Very Light (VL), 10–11 are Light (L), 12 is Slightly Moderate (SM), 13 is Moderate (M), 14 is Slightly Heavy (SH), 15–16 are Heavy (H), and over 16 are Intense (I).

The activity of stacking chairs allowed the observation of variations in the measurements of perceived effort using different methods. The results are presented in Table 5.3. It can be seen that the highest perceived exertion is increased when fuzzy logic is used. It has been found that aspects such as habituation in performing an activity can cause a person to perceive greater effort when that person

Table 5.3 Number of Users for Each Effort Level During Stacking Chair Activity, Grouped by Method

Perceived Exertion	Borg	Chamoux	Personalized Chamoux	Fuzzy Personalized Chamoux
VL	11	4	0	0
L	9	12	8	6
SM	0	4	6	6
M	0	0	4	5
SH	0	0	2	3

has no prior experience of performing that activity. And since the activity takes place at work and thus may last 8 h or more, excessive exhaustion may be perceived, despite the fact that the heart rate does not register high levels of effort.

Now, Fig. 5.6 shows a comparison of scalar results obtained by applying the Chamoux method versus our proposed method (FPC). Fig. 5.7 shows the same comparison with linguistic results. In both

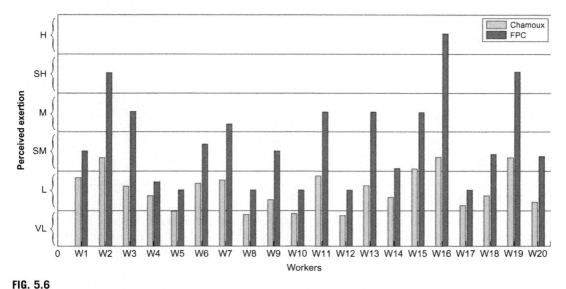

FIG. 5.6

Stacking chair activity (scalar results).

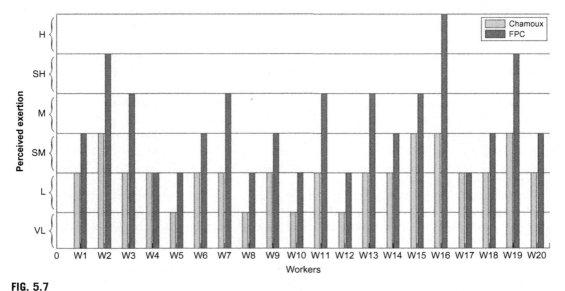

FIG. 5.7

Stacking chair activity (linguistic results).

figures, one can see that the FPC method illustrates clearly how a physical activity affects each worker. As we mentioned, Fig. 5.6 can be very useful for a decision maker to appreciate how a worker is impacted by a specific physical activity [28].

In addition to heart rate measurements, all participants were observed directly during the experiment to estimate their level of physical exertion. In addition, they were asked about their perceived effort at the end of each activity. We observed that the classification of perceived effort using our proposed method is coincident with respect to our direct observation and the responses of the participants.

The main contribution of this proposal is the ability to classify the perceived effort of people in work activities, to improve their safety and health, as it formally establishes that a person's effort can be estimated based on heart rate with adjustments according to parameters such as habituation. Effort was estimated with the standard methods for stacking chairs as a reference to analyze the gap with the personalized perceived effort estimated by our proposed method. The utility of measuring the personalized effort of workers in their work environment to lies in the preservation of their health and the possibility of determining whether people are carrying out their activities according to their personal abilities, skills and habituation for improvement of performance, safety, and welfare.

5.4 CONCLUSIONS

As established, custom monitoring to estimate heat stress is more effective and less invasive than ISO methods because it uses noninvasive sensor technology to capture the environmental parameters and workload in real time.

Our monitoring proposal can provide users with a clear vision for interpreting energy consumption and the index of the painfulness of activities to estimate heat stress at work and establish preventive actions. Some authors have validated the advantages of sensor technology for the recognition of activity and estimation of energy expenditure in workplaces.

This proposal has important advantages: it is easy to deploy, it is relatively cheap, it employs personalization, and measurements are real-time rather than table values created from a standard profile. Moreover, the equipment is ubiquitous and ubiquitously deployable; the basic infrastructure is a mobile phone, which enables self-care and timely alerting of workers to make decisions about the control of HSW based on objective information. Finally, our proposal can be used in situ for continuous monitoring of workers.

Wearing a smartphone on the hip is convenient for monitoring activities that involve a balanced use of the body and limbs. For activities that mostly make use of the hands, a sensor contained in a bracelet or wristband must be used to increase accuracy.

In situ experiments have shown that our monitoring proposal is effective in giving users sufficient data to make decisions and establish a work-rest time program, in accordance with the estimation of heat stress or work intensity.

Our solutions are good examples of how conveniently combined commercial sports and personal health monitoring devices can lead to a new generation of effective noninvasive mechanisms for monitoring health variables, in this case heat stress and perceived exertion. They also demonstrate how the awareness of context, personal sense, and mobile computation can enable AAW.

5.5 FUTURE WORKS

Personalization requires systems adaptive to the user profile, because the profile's elements can be added or removed, or can change values. Even more, the user model itself may change. Even if with context, each element can have variations in real time. Therefore, it is necessary to have mechanisms that achieve real-time adaptability of the system in terms of user model and context.

Another research line derived from this work could be to develop a health monitoring solution and its control, representing a broader approach regarding the possibility of recognition of different physiological and behavioral parameters that could affect health. The objective would be to achieve a personalized advisor system for health care.

It also would be interesting to record data over time to analyze correlations with future diseases or illnesses, or even to provide a prevention system for health monitoring.

A future evolution of this work would be to integrate itself into the Internet of Things: that is, a scenario with sensed and remotely controlled objects using the existing network infrastructure, creating opportunities for integration and exchange of information between the physical world and computer systems, leading to greater efficiency, accuracy, and economic benefits.

Regarding the case study, future work could consider bracelets containing sensors, such as skin temperature gauges or rate of sweating sensors, which would offer new opportunities. This could allow the great variety of sensors being developed to measure the influence of other factors (e.g., chemicals, noise, etc.) on the well-being of people at work.

In addition, personalized monitoring can help us to understand the effect of work adaptation on the part of individuals in the estimates obtained. We also visualize that new technologies and solutions must be developed for preventive care actions. This study highlights the need for ISO standards to consider including technological devices to build custom methods.

REFERENCES

[1] Ducatel K, Bogdanowicz M, Scapolo F, Leijten J, Burgelman JC. Ambient intelligence: from vision to reality. IST Advisory Group Draft Report, European Commission; 2003.

[2] Dobre C, Xhafa F. Intelligent services for big data science. Future Gener Comput Syst 2014;37:267–81.

[3] van den Broek G. Cavallo F, Wehrmann C. AALIANCE ambient assisted living roadmap. vol. 6. Clifton: IOS Press; 2010.

[4] Bühler C. Ambient intelligence in working environments. Universal access in human-computer interaction. Intelligent and ubiquitous interaction environments. New York: Springer; 2009. p. 143–9.

[5] Migliaccio G, Teizer J, Cheng T, Gatti U. Automatic identification of unsafe bending behavior of construction workers using real-time location sensing and physiological status monitoring. Proceedings of construction research congress; 2012. p. 633–42.

[6] Senyurek L, Hocaoglu K, Sezer B, Urhan O. Monitoring workers through wearable transceivers for improving work safety. In: 2011 IEEE 7th international symposium on intelligent signal processing (WISP). IEEE; 2011. p. 1–3.

[7] Koskimäki H, Huikari V, Siirtola P, Laurinen P, Röning J. Activity recognition using a wrist-worn inertial measurement unit: a case study for industrial assembly lines. 17th Mediterranean conference on control and automation, 2009. MED'09. IEEE; 2009. p. 401–5.

[8] International Labour Organization. Safety and health at work, http://www.ilo.org; 2015.

[9] Acampora G, Cook DJ, Rashidi P, Vasilakos AV. A survey on ambient intelligence in healthcare. Proc IEEE 2013;101(12):2470–94.

[10] Cacciagrano D, Corradini F. Healthcare tomorrow: toward self-adaptive, ubiquitous and personalized services. In: UBICOMM 2011 fifth int conf mob ubiquitous comput syst serv technol; 2011. p. 245–50.

[11] Lu YF, Goldstein DB, Angrist M, Cavalleri G. Personalized medicine and human genetic diversity. Cold Spring Harb Perspect Med 2014;4(9):A008581.

[12] Graf S, et al. Adaptivity and personalization in ubiquitous learning systems. New York: Springer; 2008.

[13] Omahony K, Liang J, Delaney K. User-centric personalization and autonomous reconfiguration across ubiquitous computing environments. In: UBICOMM 2012 sixth int conf mob ubiquitous comput syst serv technol; 2012. p. 48–53.

[14] National Disaster Management Authority, Government of India. India heat wave eases after nearly 1,700 deaths, http://www.ndma.gov.in/en/media-public-awareness/disaster/natural-disaster/heat-wave.html; 2015.

[15] Aarts EHL, de Ruyter BER. New research perspectives on ambient intelligence. JAISE 2009;1(1):5–214.

[16] Ducatel K, Bogdanowicz M, Scapolo F, Leijten J, Burgelman JC. Scenarios for ambient intelligence in 2010. Sevilla, Spain: Office for Official Publications of the European Communities; 2001.

[17] MEMS and Nanotechnologies Exchange, http://www.mems-exchange.org/MEMS/what-is.html; 2015.

[18] Beigl M, Krohn A, Zimmer T, Decker C. Typical sensors needed in ubiquitous and pervasive computing. Proc INSS; 2004. p. 22–3.

[19] Ciuti G, Ricotti L, Menciassi A, Dario P. MEMS sensor technologies for human centred applications in healthcare, physical activities, safety and environmental sensing: a review on research activities in Italy. Sensors 2015;15(3):6441–68.

[20] Ruiz-Garcia L, Lunadei L, Barreiro P, Robla I. A review of wireless sensor technologies and applications in agriculture and food industry: state of the art and current trends. Sensors 2009;9(6):4728–50.

[21] Garcia-Ceja E, Brena RF, Carrasco-Jimenez JC, Garrido L. Long-term activity recognition from wristwatch accelerometer data. Sensors 2014;14(12):22500–24.

[22] Attal F, Mohammed S, Dedabrishvili M, Chamroukhi F, Oukhellou L, Amirat Y. Physical human activity recognition using wearable sensors. Sensors 2015;15(12):31314–38.

[23] Reinert D, Flaspöler E, Hauke A, Brun E. Identification of emerging occupational safety and health risks. Saf Sci Monit 2007;11(3):3.

[24] Doe R. International Labour Standards on occupational safety and health, http://www.ilo.org/; 2013.

[25] Zato C, Sánchez A, Villarrubia G, Bajo J, Rodríguez S, Paz JFD. Personalization of the workplace through a proximity detection system using user's profiles. In: 7th international conference on knowledge management in organizations: service and cloud computing. Springer; 2013. p. 505–13.

[26] World Health Organization. Occupational health, http://www.who.int/occupational_health/topics/stressatwp/en/; 2015.

[27] Pancardo P, Acosta FD, Hernndez-Nolasco JA, Wister MA, de Ipia DL. Real-time personalized monitoring to estimate occupational heat stress in ambient assisted working. Sensors 2015;15(7):16956–80.

[28] Pancardo P, Hernández-Nolasco JA, Acosta FD, Wister MA. Personalizing physical effort estimation in workplaces using a wearable heart rate sensor. Cham: Springer International Publishing; 2016. p. 111–22.

[29] Memon M, Wagner SR, Pedersen CF, Beevi FHA, Hansen FO. Ambient assisted living healthcare frameworks, platforms, standards and quality attributes. Sensors 2014;14(3):4312–41.

[30] Cho J, Kim J, Kim T. Smart phone-based human activity classification and energy expenditure generation in building environments. In: Proceedings of the 7th international symposium on sustainable healthy buildings, vol. 2012; 2012. p. 97–2105.

[31] Bouten CV, Westerterp KR, Verduin M, Janssen JD. Assessment of energy expenditure for physical activity using a triaxial accelerometer. Age (yr) 1994;23:21–7.

[32] Noble BJ, Robertson RJ. Perceived exertion. Illinois, United States: Human Kinetics Publishers; 1996.

[33] Ergonomics guide to assessment of metabolic and cardiac costs of physical work. Am Ind Hyg Assoc J 1971;32(8):560–4. http://www.tandfonline.com/doi/abs/10.1080/0002889718506506.

[34] Medjahed H, Istrate D, Boudy J, Baldinger JL, Bougueroua L, Dhouib MA, et al. A fuzzy logic approach for remote healthcare monitoring by learning and recognizing human activities of daily living. Fuzzy logic-emerging technologies and applications; 2012. InTech.

[35] Zadeh LA. Outline of a new approach to the analysis of complex systems and decision processes. IEEE Trans Syst Man Cybern 1973;1100:38–45.

[36] Bouton ME. Learning and behavior: a contemporary synthesis. England: Sinauer Associates; 2007.

[37] Borresen J, Lambert MI. The quantification of training load, the training response and the effect on performance. Sports Med 2009;39:775–9.

[38] Heath EM, Blackwell JR, Baker UC, Smith DR, Kornatz KW. Backward walking practice decreases oxygen uptake, heart rate and ratings of perceived exertion. Phys Ther Sport 2001;2(4):171–7.

[39] Frimat P, Amphoux M, Chamoux A. Interprétation et mesure de la fréquence cardiaque. Revue de Medicine du Travail XV 1988;4:147–65.

HOME AUTOMATION ARCHITECTURE FOR COMFORT, SECURITY, AND RESOURCE SAVINGS

6

Armando García B., Erica Ruiz I., Joaquín Cortez G., Adolfo Espinoza I.
Sonora Institute of Technology (ITSON), Ciudad Obregón, México

6.1 INTRODUCTION

Historically, technological developments have benefited society, and this has motivated researchers to continue working on current technological problems by generating alternative solutions. In recent years, due to the increase in the population worldwide, resources such as water and electricity have become increasingly scarce and difficult to generate. On the other hand, solar energy is a resource that has not been used as it should, and which represents a huge contrast in regions where there is no electricity but lots of solar radiation. Today, more than ever, it is necessary to apply technology to optimize the resources that man uses every day and that are wasted for different reasons. This is increasingly important in large cities where the number of homes is growing, demanding more and more services, so that new schemes are needed to optimize resources.

Home automation did not really appear until the 1970s, with the advances in technology and services at the time renewing perspectives on the modern house [1]. Based on the principle of automation, home automation appears as an alternative to solving problems due to the absence of human beings in some situations where resources are not properly used or are wasted. Home automation refers to the integration of information technology, automation, and communications with the aim of providing better quality of life, security and comfort to home occupants, and additionally making better use of the available resources [2,3].

Today's modern houses are visualized as those in which the main activities are fully automated, providing a higher degree of comfort, and which also exhibit the special characteristic of optimizing resources. This is possible thanks to developments in electronics, wireless communications, sensor networks and actuators, and the recent Internet of Things (IoT) technologies. A large variety of actuator sensors now allow monitoring and control of any process within a home. This opens a niche of opportunity to fulfill the role that has been assigned to home automation: to provide security and comfort, and to contribute to the efficient use of resources [4,5].

Intelligent Data Sensing and Processing for Health and Well-being Applications. https://doi.org/10.1016/B978-0-12-812130-6.00006-8

6.1.1 **RELATED WORK**

In the context of home automation based on wireless communications technology, different proposals can be seen in the literature. In this section we discuss the most relevant to our work. As far as we know, Nunes and Delgado [3] developed one of the first architectures proposed for home automation. The architecture is wired and uses the RS-232 or EIA-485 protocol to communicate with the modules. The article introduces the concept of supervising and control modules. Gill et al. also proposed a home automation system using Zigbee and wi-fi technologies [2]. The solution consists of a network based on Zigbee linked to the home automation system (virtual home) and a gateway to communicate with remote users. The research does not show results for the interface with the end user. The author of [4] introduces the use of Android-based smartphones and tablets to interact with the user outside of the home through web communications. Although it is possible to control several sensors inside the house, it does not use indoor wireless communication; instead, an Arduino Ethernet server is implemented.

According to Yang et al. [6], a solution to implement the Zigbee and 802.15.4 protocols in a single chip (CC2430) is possible. However, these authors focus only on the results of the communication between the gateway and coordinator and the end device and the sensors. The article does not provide results for the control of the sensors and user interface. A different approach using nodes based on PIC is studied by Alkar et al. and Mai et al. [7,8]. Both studies emphasize the implementation of the receiving and transmitting unit, presenting results of the implementation for both controls. The systems are used to turn on lamps with a graphical interface. In the first one, the interface is web based, while the second one is C# based. To improve secured access, the author of [5] developed a wireless home automation system. The main contribution is the speech recognition and the support for elderly people. A design for home automation is also presented by Zhihua [9], based on a wireless sensor network. A PC machine is used as a gateway and home control center. In spite of all the elements considered in the automation system, the authors do not present results for the implementation and tests for the overall system communication. In recent studies, versions of Zigbee-based home automation are also presented [10,11]. The articles focus on one or two variables (humidity and temperature), which is too restricted for wider applications in which it is necessary to monitor diverse activities.

According to Suryadevara and Mukhopadhyay [12], a solution using a home monitoring system can provide enhanced wellness for the elderly. The approach is well oriented towards the implementation of sensing units for different variables, such as temperature, humidity, and light intensity. The action of monitoring is well defined; however, the security requirements are not clearly defined or considered. In Ransing and Rajput [13] a smart home for elder care, based on a wireless sensor network, is presented. The contribution is a LabVIEW user interface to monitor several parameters, as well as the sending of an SMS message in case of hazards. However, only the temperature is shown in the interface.

An extension of the research presented by Suryadevara in Ref. [12] is proposed in Ref. [14] in order to diminish costs, increase real-time characteristics and have better power management. The problem of interference present in Zigbee-based smart home designs is addressed in Ghayvat et al. [15]. This factor is critical for certain situations, especially when security issues are considered. Several tests were performed to investigate the optimal distance between transmitter and receiver using PSR and RSSI values. Tests for channel interference were carried out to investigate the performance in the presence of wi-fi channels. However, the authors are not interested in aspects such as the architecture and end-user interface. Additional characteristics and functionalities are added to the smart home monitoring system considered by Ghayvat et al. [16]. The approach analyzes different parameters such as interference and attenuation. They perform a number of tests for the monitoring system in the presence of technologies such as wi-fi, Bluetooth and microwave. They stress the importance of wellness. On the other hand, the IoT and smart building concepts are linked to the wireless sensor network approach.

The most challenging problem with the previously mentioned proposals is to integrate the most important characteristics and functionalities of smart home automation, considering flexibility, low cost, and expansion. The main contribution of our work consists of an architecture that can take into account comfort, security, and resource saving in one single platform. As we have seen, the Zigbee protocol is a mature technology used extensively, and for that reason it is the base for our architecture.

The remainder of the chapter is organized as follows: first we describe the methodology of the proposed architecture, as well as the considered subsystems; next we expose the design for each element of the system divided into elements of the network; then we present the results obtained from the tests and validation; and. Finally, the conclusions from the results are analyzed.

6.2 METHODOLOGY AND IMPLEMENTATION

In this section, the architecture to control the different subsystems (the term subsystem in this chapter is used to differentiate from the complete main system) that are in an average house and the design of their main elements are explained. The goals of the whole system are to optimize electrical energy and water usage and take advantage of solar energy when available, as well as providing security in risky situations. The subsystems considered in an average home are

- Lighting control
- Shutter control
- Irrigation control
- Detection of gas leakage

Lighting control. This subsystem is responsible for controlling all the lights both inside and outside the house. Because the lighting is located in different areas, it needs to be segmented. The control for all the lights will be on and off. It is also intended to control the lights that are on and those not being used, for energy-saving purposes.

Shutter control. The curtains/shutters located in each of the rooms are the responsibility of this subsystem. Besides controlling the shutters, another objective is to take advantage of the natural light during the day by opening curtains, and to close them at night. To achieve this, coordination with the lighting control subsystem is required.

Irrigation control. In order to provide comfort as well as economize on water resources, the irrigation subsystem is integrated. This system must have the necessary intelligence to determine when to start the irrigation system according to the level of humidity in the gardens or lawn. Likewise, this subsystem must detect possible climatic conditions in order to select the appropriate time to start the irrigation process, as well as control its duration.

Detection of gas leakage. A feature of home automation is the need for elements that reduce risk situations. The gas detection subsystem provides a condition for detecting the presence of leaks in the connections or the heater. In this way the inhabitants are made aware of situations that are dangerous.

Three aspects are fundamental to home automation: comfort, conserving resources, and security.

Comfort is provided by automated processes where the user does not need to move or take action to perform a control action for the devices or subsystems in the home. However, the option does exist to be able to choose the manual mode. If someone is in a specific area, they have the option to perform the control there manually.

Conserving resources. The proposed system considers the importance of energy and water expenditure. By having control over the lights, shutters, and irrigation system, the system avoids wasting electrical energy and water. In addition, gas savings are considered when detecting any leaks present in the connections.

The security issue is an important aspect of the system. In this chapter, safety refers to reducing risks for the inhabitants of both the exterior and the interior. As already mentioned, gas detection avoids the possibility of accidents that may occur. As shown in the following text, the system is flexible since it allows additional subsystems, such as railings, open doors, and other devices.

6.2.1 PROPOSED ARCHITECTURE

In this section, the organization of the architecture in order to achieve the goals mentioned is explained. The platform comprises a brain that is responsible for coordinating all the activities in the system as well as interacting with the end user. The second element in the architecture is the network for communication with the participating nodes, and, finally, the remote nodes in which the different subsystems can reside, as shown in Fig. 6.1. The network consists of a coordinating module and a series of remote modules based on a microcontroller (MCU) and an Xbee RF transceiver from Digi International. The coordinating module, as the name implies, is responsible for coordinating actions on the network. The remote modules are responsible for each of the activities to be controlled in the home, such as lighting, shutters, temperature, etc. The architecture has flexibility, allowing it to be increased (or deleted) by any activity simply by adding a remote module and the respective sensors and actuators required for each case. In both nodes, coordinator and remote, the Zigbee Alliance protocol stack [17] and the IEEE 802.15.4 standard were used. As seen in the same figure, diverse remote modules can control different subsystems.

6.2.2 DESIGN AND IMPLEMENTATION

The design was based on four elements: (1) the brain of the system, (2) the network topology, (3) the coordinator module, and (4) the remote modules. Here we describe these elements and their functionalities as they were designed, as well as the characteristics of the main subsystems: shutter, illumination, irrigation, and gas control.

The system's PC or the brain provides a graphical user interface through which it can interact with remote subsystems. Using a Communications Device Class (CDC) USB connection to the coordinator node MCU, parameters are requested and modified that can be used by applications on remote MCUs, as shown in Fig. 6.2. To detect remote nodes within the scope range of the coordinator, the system has the option to perform a test and require the transmission power to identify the remote devices that make up the network.

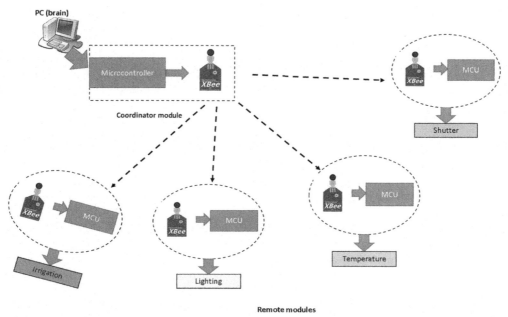

FIG. 6.1

Diagram of the proposed architecture.

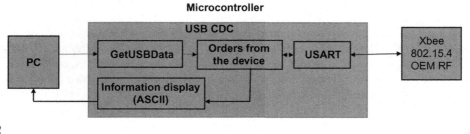

FIG. 6.2

Communication of the PC with the coordinating node.

6.2.2.1 Network topology

As already mentioned, ZigBee is used for communication [7–13]. The topology that conforms to the proposed platform is the star topology, in which the network is controlled by a single device called the ZigBee Coordinator (ZB). This device is responsible for starting and maintaining the devices on the network. All other devices, known as end devices (ZED, ZigBee End Devices), establish direct communication with the coordinator. The elements that are used as coordinator and the final devices are the Xbee modules, which are connected with a low-cost microcontroller, in charge of realizing all the logic of the modules.

The coordinating module is composed of a main MCU connected to an Xbee OEM radio frequency (RF) module. In the same way as in Alkar et al. [7] and Mai et al. [8], the design of the module is based on PICs in the transmitting and receiving nodes. The function of the module is to interpret the commands sent by the user from the PC, as well as to receive the status of each of the components of the system, to display them in the graphical interface. Communication between the two devices is done using the API mode of operation of the XBee modules. The MCU coordinator is in charge of configuring the RF module to initiate as network coordinator, as well as to keep the information of the remote devices, such as the physical address and device number stored. Another function of the coordinating MCU is to resolve association messages sent by the remote devices. According to the list of devices housed in its memory, the MCU responds by indicating the device's assigned number, retained or modified by the remote devices. In this way, conflicts in the network are avoided and the user interface can reference them in a simple and effective way, avoiding the identifiers changing between cycles of energy. The coordinating MCU stores the physical addresses of the remote devices in nonvolatile memory.

The remote module is responsible for controlling one or more of the activities to be automated in the home. Its composition is similar to those of the coordinating module, an MCU, and an Xbee RF module. The difference between this and the other nodes lies in the configuration, since it is configured as an end device. Just as the coordinating module uses the API operating mode to configure and to establish communication through its respective Xbee RF module, the remote MCU manages the operation mode of its RF module and interprets the commands sent from the coordinator module. The remote module has a task to establish communication with its coordinator each time it is energized; in this way it makes a request of association where the coordinating MCU resolves a number of devices for its operation. This parameter is stored in the Xbee RF module of the remote device. Only the coordinator decides whether the remote device operates with the same number or if it is necessary to assign a new one. The association request also has the function of communicating the physical address to the remote device. After the remote device is started, it performs an active scan, looking for all the nodes around it. Once the response is received, the remote module operates on the parameters of the coordinator: Channel and PAN ID (Network Identifier). In this way the brain knows the number of remote modules in the network corresponding to the automated activities in the house. For both types of modules, coordinator and remote, communication with the RF modules is established via the USART port, using the interrupt mode on the MCU. The software in the remote modules consists of two parts. One is the process control software, which performs the specific operation logic of the module. The other is the communication software, this works according to interruptions generated by the serial port, each time a command from the Xbee transceiver is received. It should be mentioned that each remote module can have different subsystem, for example in an MCU can be irrigation, lighting (outside in this case) and the shutters of any of the rooms adjoining the irrigation area, as shown in Fig. 6.3.

The brain PC is responsible for monitoring each of the subsystems existing in the house, in this case irrigation, lighting, and shutters. To do this, it is necessary to communicate with all the remote nodes. Each remote node makes use of the scheme shown in Fig. 6.3. For each particular case, the inputs and outputs of the PIC or MCUs are used for a given subsystem. An example of this is shown in Table 6.1.

Once the function is executed for each subsystem, the data is sent to the memory location with the information of the task performance and the state it is in. Subsequently, the MCU of the remote module sends this information to the coordinating module through the Zigbee communication to notify the PC. To execute a task, each remote module contains a "scheduler" or time control. This is responsible for

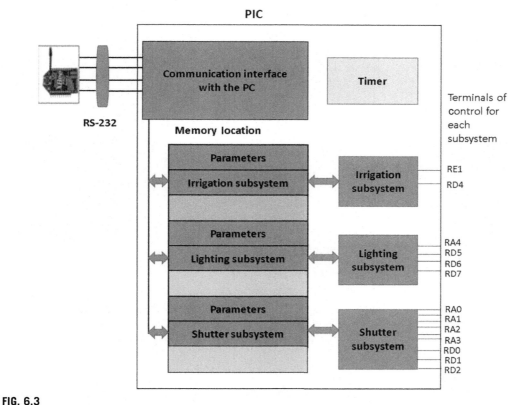

FIG. 6.3

Remote module with irrigation, lighting, and shutter subsystems.

Table 6.1 Control Terminals That Correspond to Each Subsystem

Subsystem	Input	Output
Illumination	RA4	RD5, RD6, RD7
Shutters	RA0, RA1, RA2, RA3	RD0, RD1, RD2
Irrigation	RE1	RD4

controlling the working time of each subsystem by means of periodic interruptions. The operation of the timer is as follows: every time the timer interrupts, the service routine increments a counter and attends to one of the subsystems, depending on the value of the counter. This is reset to 0 when it has all been serviced. To carry out the tests to each of the remote modules, an LCD display was added, which can optionally be removed in the operating mode. The following sections describe each of the subsystems that the platform can control.

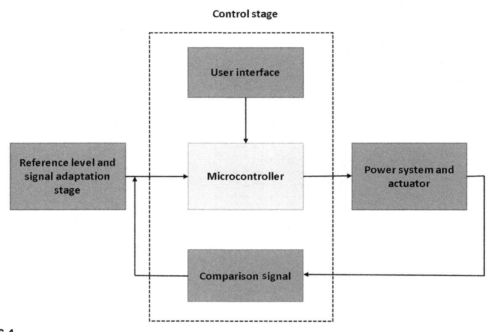

Control stage

FIG. 6.4

Block diagram of the shutter control.

6.2.2.1.1 Shutter subsystem

The subsystem consists of three stages: signal adequacy, control, and power, as shown in Fig. 6.4.

The adaptation of the signal is carried out with a photoconductive cell whose function is to refer to the state in which the control is located, perceiving the intensity of the sun that passes through the shutter. In addition, a Wheatstone bridge is implemented as a coupling circuit of the photoconductive cell.

On the other hand the *control stage* is formed by a user interface, a microcontroller, and a comparison signal. The user interface operates via interrupts, either from a matrix keyboard or a timer, whose function is to start summer mode in the event of no information from the keyboard. A liquid crystal display is used for the data display of the microcontroller. When the circuit is turned on, a start message is displayed, and then the automatic or manual operation mode is selected. Fig. 6.5 shows the messages displayed:

The PIC16F877A *microcontroller* is used as the base of the data acquisition module, which interprets the commands from the keyboard for the control and allows the visualization of the variables through the LCD. The microcontroller receives two signals from the ADC, the signal from the conductive photocell and the signal from the potentiometer, which are compared to each other to determine the desired position of the shade. The system can be controlled automatically or manually. The automatic control has two modes of programming: winter and summer. In winter, when the conductive photocell marks greater light, is when the shutter must be opened more to raise the temperature of the room.

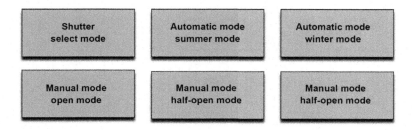

FIG. 6.5

LCD messages.

In contrast, in summer, it is recommended to avoid the rays of the sun. In this mode, the controller compares the input signal from the conductive photocell with the potentiometer. The comparisons are different, depending on whether it is winter or summer. Port C generates the three control bits, showing the following combinations: 000 for the engine off, 101 the engine rotates to the right by closing the shutter, and 110 the engine rotates to the left by opening the shutter. The manual control compares the state in which the shutter is located with the potentiometer with the current data entered, taking into account that the potentiometer does not rotate 360 degrees. The three modes that are handled in the project are closed shutter 0 degree, half open 45 degrees, and open 90 degrees.

The control program is loaded into the microcontroller and is based on assembler language, referenced from the general system design. For this purpose, the MPLAB program was used as the compiler of the aforementioned language. Once the system is initialized, it loads some variables that are used to configure the PIC, and then a "BEGINNER: SELECT FORM" message is displayed on the LCD and the program waits until a key is pressed. In the case of selecting a form of control and assigning the variable, the initial program will recognize the variable and it will go to the assigned routine. The routines of the two forms of control are different and cover the different needs of each; moving the direct current motor according to the design described above each key has different routines, but they always have in common that all use the ADC of the PIC. In the case of automatic control, the two input signals are read in the analog to digital conversion and based on this reading a comparison of the two signals is made. Otherwise, in manual control, only the input signal of the potentiometer is read and is compared to a predetermined variable. After performing the comparison based on each of the routines as described previously, the routine displays three bits depending on which was higher or lower. The comparison signal corresponds to a potentiometer and is part of the shutter control, since this is a signal that indicates the feedback as to the state of the mechanical shutter system. The potentiometer varies its resistance depending on how the cursor is moved that is attached to the shade rod. When the shade hatches move, the slider changes the resistance and throws a voltage.

In the *power stage* we worked with the basic circuitry for proper operation of the CD engine. Devices such as diodes, TIP31C and 32C allow the forming of an H bridge, with which they can perform the operation of a motor. In this case the management is bidirectional, with rapid braking and with the possibility of easily implementing speed control. The outputs have a design that allows direct handling of DC motors. Fig. 6.6 shows the complete design in which the three stages are integrated.

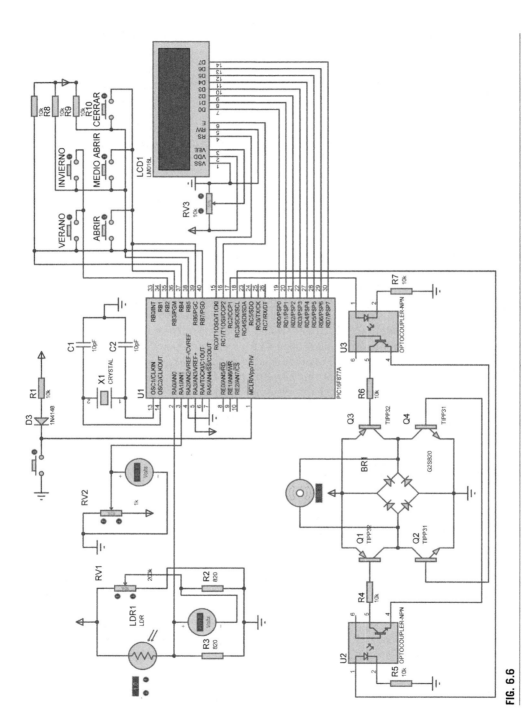

FIG. 6.6

ISIS simulation of the shutter control system.

6.2.2.1.2 Irrigation control subsystem

As a background to the topic of irrigation, in general in Mexico most of the water consumed is dedicated to agriculture, 75.72%, to irrigate some 5,400,000 ha. The rest, 24.28%, is used in industries and households. The use of water has increased in recent years for domestic use. There is no exact data on how much water is lost due to leaks or misuse. Therefore it is essential to make an appropriate use of water resources.

In order to contribute to the improvement of the irrigation process, an irrigation subsystem control is proposed. As mentioned previously, this process is controlled with a remote module, the composition of which has already been described. For the particular case of the irrigation system, the Honeywell HIH-001 moisture sensor was used. This sensor works with a power supply voltage of 4.5–5.8 Vdc, making it compatible with the system power. The sensor has the advantage of being linear and of connecting directly to the analog input of the microcontroller. In addition to the humidity sensor, it requires spray equipment that is activated by a relay and an electrovalve. The user, through a menu, has the option to choose the manual mode or automatic mode as shown in Fig. 6.7.

Based on the studies of soil moisture, 50% and 80% moisture were taken as the reference for the limit values for the igniter and shutdown of the sprinkler. The control of irrigation of the garden has to comply with both manual and automatic operation.

- *Manual Option*. If the user wants to change the state to manual operation of the water sprinklers, a coordinating module signal will be sent to update the status via the interface.
- *Automatic option (via sensors)*. There are two ranges of humidity: the high level and the low level. If the sensor detects the low level, a relay will be activated, which in turn, turns on a solenoid valve and a sprinkler to irrigate the garden. If the sensor detects the high level, it will deactivate the relay; therefore, the irrigation will be stopped, until the sensor detects the low level again.

6.2.2.1.3 Illumination control subsystem

The control of home lighting is one of the most fundamental requirements in home automation. By automating the lighting, there is a considerable savings of electrical energy consumption. This system allows the lighting to be turned on only when there is presence of one or more people in a certain room. In addition, energy can be saved by controlling the intensity of the light source so that it is only what is needed, depending on natural lighting. Another advantage of these systems is that they can be used for safety. They allow the user to turn on the different light sources of the house, either in a programmed or random manner, to simulate the presence of people and to give the impression that the house is

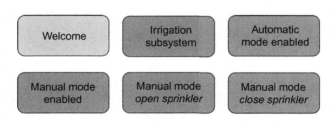

FIG. 6.7

Menu to select the working mode of the irrigation subsystem.

inhabited. To achieve the aforementioned functions, it is necessary for each lighting source to have an actuator that controls the power. In addition, the subsystem must also have sensors for the presence of people and sensors of light intensity.

For the physical implementation of this subsystem, the module is required to dedicate a one-bit digital input for the presence sensor, an analog input for the lighting sensor, or a one-bit digital input handling PWM signals. For the control output of the luminaire, a PWM output and power switching circuitry are required.

6.2.2.1.4 Gas control subsystem

This subsystem detects whether the gas is open, and whether or not there is a flame in the burners. It is able to close or open the gas if necessary, to avoid an accident. Like the other subsystems, a series of sensor and actuator arrangements are used with a control stage. In the first stage, two thermocouples are used to detect whether there is a flame in the burners. Thermocouples are devices formed by the union of two different metals which produce a voltage (Seebeck effect) when varying the temperature. The thermocouples' output voltage was then amplified from 3 mV to 4 V with an arrangement of operational amplifiers, since the control stage operates with logic levels. The arrangement was designed to reach the point of saturation with the smallest flame and in the shortest time. This creates a kind of on/off switch with a small delay. Subsequently, using a PIC, a controller was designed to be able to detect the inputs and decide if it is necessary to open or close the gas, to mark an error in one of the burners or to activate the ignition system.

Using a solenoid valve, a mechanism capable of opening or closing the gas by an electric pulse was designed. In addition, a system of ignition for the burners was designed that works with alternating current of 110 V; a MOC3010 (optocoupler) was used to activate it with the proposed control system. Finally, two LEDs were used to indicate the errors in the different burners. Fig. 6.8 shows the general design of the gas control subsystem and Fig. 6.9 shows its simulation.

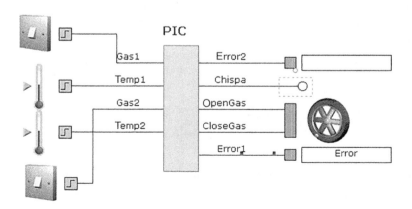

FIG. 6.8

General gas control design.

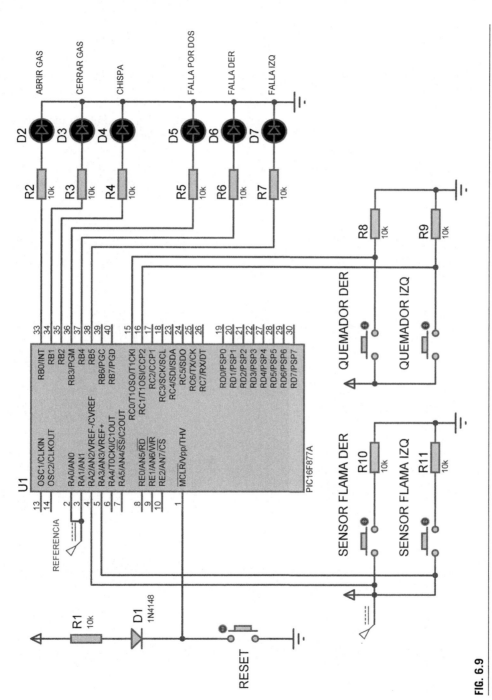

FIG. 6.9

ISIS simulation of the gas control subsystem.

6.3 RESULTS OF THE INTEGRATED HOME AUTOMATION SYSTEM

In this section, the results obtained from the tests are presented. Once each of the stages has been integrated (PC or brain, communication network, coordinating module, and remote modules), the final domotic system is obtained, with each of the subsystems as shown in Fig. 6.10. As can be seen, each remote module can contain five subsystems at most. The network topology used is star, with only one coordinating node. If required, there may be two or more coordinators, depending on the area to be covered in the dwelling. In this case a single coordinating module is presented. The subsystems to be integrated will depend on their location in or outside the house. For example, in a room it is possible to have the lighting, shutter, and temperature control subsystems. The exterior of the house can have irrigation, and lighting if people are near the garden. The gas control subsystem can be separate from the others.

6.3.1 VALIDATION OF THE SYSTEM

A number of tests were performed to validate the operation of the system. Communication between Xbee modules used the X-CTU software provided by the manufacturer.

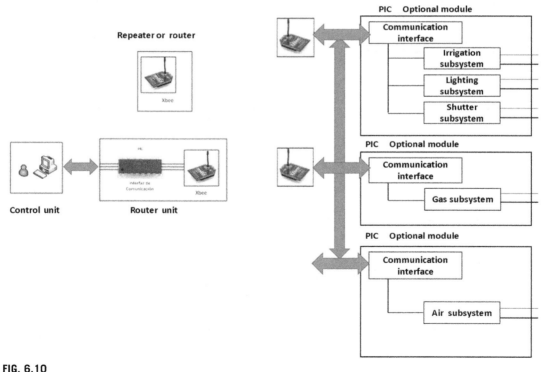

FIG. 6.10

Integrated home automation system.

6.3.1.1 Coordinator mode settings

Using the X-CTU software's modem configuration window, we can view the values of the main configuration parameters within the coordinator module. These registers and their respective factory values are shown in Fig. 6.11.

As a first step, communication with the MCU is performed; this determines which Xbee module will act as a coordinator. The PAN ID and the High and Low serial number (SH and SL) are set and unchanged. The module (*DomoticaCoordinador*) has an assigned name. Once the communication is established, the Xbee module operates as a coordinator, as shown in Fig. 6.11.

The next step is to enable the API mode in the coordinator. This is achieved by an MCU command of the module. The result is shown in Fig. 6.12.

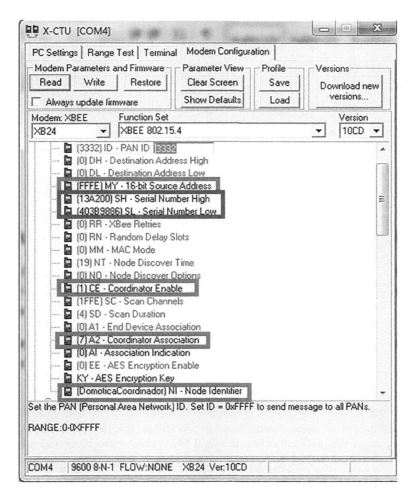

FIG. 6.11

Xbee module configuration as coordinator.

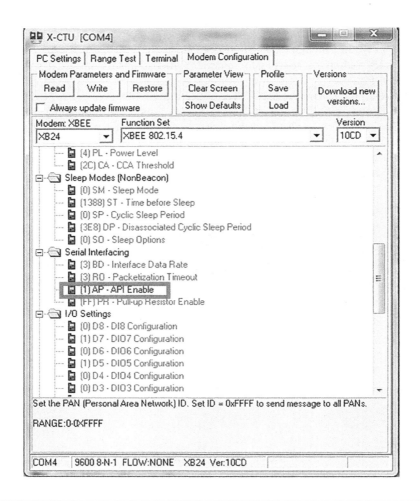

FIG. 6.12

Activation of the API mode in the coordinator module.

6.3.1.2 Association of remote modules

In order to form the network, tests were performed to associate the remote modules with the co-ordinator, by reading the parameters stored in both modules. For the device association process, one of the objectives of the platform is that the final MCU be able to locate and associate with the network parameters chosen by the XBee coordinator module, so that it starts with the reading of the parameters in X-CTU of the 802.15.4 devices arranged in this network. It should be noted that, for this example, it was decided to restore the remote module to its factory values to better show

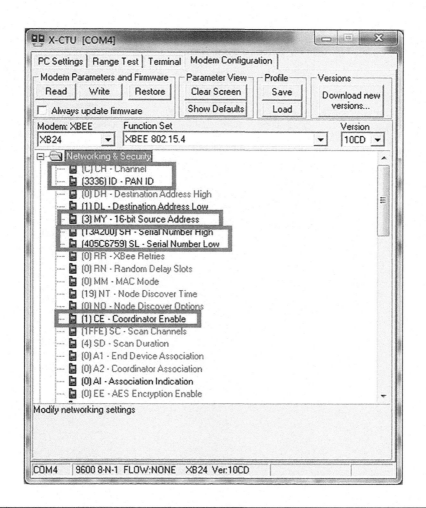

FIG. 6.13

Network parameters obtained from the remote Xbee module before being configured.

how the microcontroller acts on that module. Fig. 6.13 shows the network parameters obtained from the Xbee module that is connected to the microcontroller in charge of managing the remote device.

It should be appreciated that the parameters shown in Fig. 6.13 do not coincide with the network parameters that are obtained from the configuration of the module. This module is configured to operate as a coordinator, in 16-bit addressing mode, as well as operating under a different PAN ID. The connection to the RF module is made to the implemented circuit to read the changes in the parameters for the remote RF device again. Once the microcontroller of the remote module is connected, the parameters shown in Fig. 6.14 are obtained and coincide with the coordinator module.

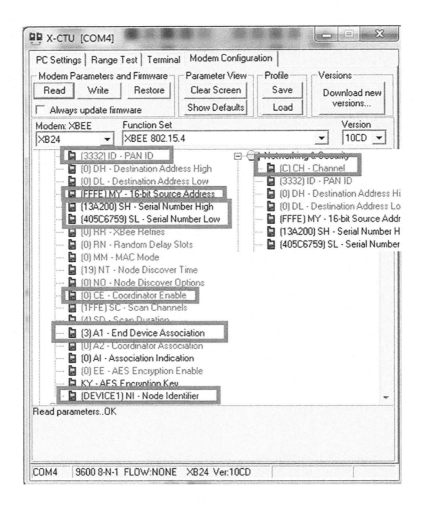

FIG. 6.14

Parameters that are obtained once the microcontroller of the corresponding remote module is connected.

The addressing mode changes to 64 bits as expected and it is also possible to see that the NI parameter changes its value to "DEVICE 1" corresponding to the device number assigned by the coordinating MCU. The described association process is performed for each of the remote modules required to cover the dwelling, so that the final network is formed. In this case the test was performed with three remote modules.

Finally, to validate the operation of the domotic system and its interaction with the user, an interface was made in Visual C#, as shown in Fig. 6.11. Because it was not possible to install the system in a real house, Prototype level for each of the subsystems was used, in addition to sending control data to verify their operation. The main controls for the interface are presented in Fig. 6.15.

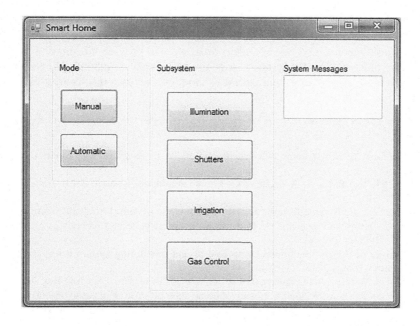

FIG. 6.15

Graphical interface for interacting with the subsystems.

6.4 **CONCLUSIONS**

Home automation is a challenging activity when we try to provide comfort, resource savings, and security. Thanks to new services and technologies, it is possible to perform difficult tasks using new communication networks, powerful microcontrollers, smart sensors and actuators. In this research, a novel architecture based on Zigbee was presented. We focused our design on four subsystems present in daily activities of an average home: control of illumination, irrigation, shutters, and gas leakage detection. We presented the implementation of a coordinator and remote modules based on a PIC and RF transceiver to perform the Zigbee and 802.15.4 protocols, as well as the procedure required to enable them. We prove the communication in the network, sending and receiving valid data. We tested three remote modules and the communications with the brain. We used a friendly visual interface for the end user developed in Visual C#, to validate functionalities of each subsystem. The results of controlling the subsystems by the brain and allowing an optional manual mode show that the comfort characteristic is provided. In the same way, controlling the time length of lights and curtain/shutter operation and optimizing the irrigation systems help to demonstrate the resource savings functionality. Finally, security is addressed by monitoring gas leakage detection.

Other key characteristics of the proposed system are the flexibility and low cost. Flexibility is provided by allowing the option to add new remote modules as they are required. Additionally, as we mentioned, it is possible to implement one, two, or four subsystems in each remote node. In this way, it is possible to provide full coverage for a house.

In the near future, we will install our system in a real house in order to prove out the communication in a real scenario, as well as to perform tests to investigate battery life for the remote modules in normal conditions based on the consumed power. Afterwards, we will provide information to determine the average size of the house where our system can be used. We also are planning to perform tests to investigate the resources saved over a long period of time and increase the security functionality.

REFERENCES

[1] Cyril JA, Malekian R. Smart home automation security: a literature review. Smart Comput Rev 2015; 5(4):269–85.

[2] Gill K, Yang SH, Yao F, Lu X. A ZigBee-based home automation system. IEEE Trans Consum Electron 2009;55(2):422–30.

[3] Nunes R, Delgado J. In: An architecture for a home automation system. 1998 IEEE international conference on electronics, circuits and systems. Surfing the waves of science and technology (Cat. No.98EX196), Lisbon. vol. 1; 1998. p. 259–62.

[4] Piyare R. Internet of things: ubiquitous home control and monitoring system using android based smart phone. Int J Intern Things 2013;2(1):5–11.

[5] Benjamin Arul S. Wireless home automation system using ZigBee. Int J Sci Eng Res 2014;5(12):133–38.

[6] Yang L, Ji M, Gao Z, Zhang W, Guo T. Design of home automation system based on ZigBee wireless sensor network. 2009 first international conference on information science and engineering, Nanjing; 2009. p. 2610–3.

[7] Alkar AZ, Gecim HS, Guney M. Web based ZigBee enabled home automation system. 2010 13th international conference on network-based information systems, Takayama; 2010. p. 290–6.

[8] Mai L, Oo MZ. Design and construction of microcontroller based wireless remote controlled industrial electrical appliances using ZigBee technology. Int J Sci Res Eng Technol 2014;3(1):79–84.

[9] Zhihua S. Design of smart home system based on ZigBee. 2016 international conference on robots & intelligent system (ICRIS), Zhangjiajie; 2016. p. 167–70.

[10] Al-Ali AR, Qasaimeh M, Al-Mardini M, Radder S, Zualkernan IA. In: ZigBee-based irrigation system for home gardens. Communications, signal processing, and their applications (ICCSPA), 2015 international conference on, Sharjah; 2015. p. 1–5.

[11] Rerkratn A, Kaewpoonsuk A. In: ZigBee based wireless temperature monitoring system for shrimp farm. 2015 15th international conference on control, automation and systems (ICCAS), Busan; 2015. p. 428–31.

[12] Suryadevara NK, Mukhopadhyay SC. Wireless sensor network based home monitoring system for wellness determination of elderly. IEEE Sensors J 2012;12(6):1965–72.

[13] Ransing RS, Rajput M. In: Smart home for elderly care, based on wireless sensor network. 2015 international conference on nascent technologies in the engineering field (ICNTE), Navi Mumbai; 2015. p. 1–5.

[14] Suryadevara NK, Mukhopadhyay SC, Kelly SDT, Gill SPS. WSN-based smart sensors and actuator for power management in intelligent buildings. IEEE/ASME Trans Mechatron 2015;20(2):564–71.

[15] Ghayvat H, Mukhopadhyay SC, Gui X. In: Addressing interference issues in a WSN based smart home for ambient assisted living. 2015 IEEE 10th conference on industrial electronics and applications (ICIEA), Auckland; 2015. p. 1661–6.

[16] Ghayvat H, Mukhopadhyay S, Gui X, Suryadevara N. WSN- and IOT-based smart homes and their extension to smart buildings. Sensors 2015;15:10350–79.

[17] ZigBee Alliance. ZigBee specifications, http://www.zigbee.org/zigbee-for-developers/network-specifications/zigbeerf4ce/.

SECURITY, PRIVACY, AND ETHICAL ISSUES IN SMART SENSOR HEALTH AND WELL-BEING APPLICATIONS

Jan Sliwa

Bern University of Applied Sciences, Biel, Switzerland

7.1 INTRODUCTION

We currently face immense growth of pervasive systems and applications gathering information about the real world and acting on it, commonly called the Internet of Things (IoT) or Internet of Everything (IoE). In this landscape a special role is played by health and well-being applications. Other IoT applications can also interact intensively with human lives—a Smart Home can lock the doors if a strange behavior of the owner is detected.

However, health applications interact directly with the most personal item, the human body. Therefore security is of utmost importance. As in processes where vital data are collected, there is a risk of privacy breach if those data are not adequately protected. Obviously, using such invasive devices raises ethical problems.

In this chapter we will discuss those issues. It is a complex area, as with the flood of novel devices we observe overlapping of the domains of the commercial market and professional healthcare, with different value systems and business models. The danger of a malicious action is only a part of our discussion, as the risk of inadvertent damage caused by hasty design, insufficient testing, omitting special conditions, inadequate skills, or deriving false conclusions from worthless data is equally important.

We will rather present the problems than give ready-to-use solutions. Due to the extremely multidisciplinary nature of the area it seems important to make the reader aware of various issues that are not evident from any particular point of view, for example, for software developers. The rapid changes make specific solutions soon obsolete. Also limited space does not permit a very detailed discussion. We also elaborate on ethical issues, as the domain of sensor-based smart medical devices goes well beyond mere technology.

7.2 STATE OF THE ART

As an introduction to the problems of the future of medicine, the discussion at the World Economic Forum in 2017 in Davos[1] can be recommended. Atul Gawande, renowned surgeon and author of *Being Mortal*, and Francis Collins, former leader of the Human Genome Project, discussed the future of healthcare leadership and biomedical research. A major trend is the change from rescue medicine to lifelong incremental care, as well as data-driven healthcare.

Good overviews of the state of the art in the area of smart medical devices can be found in [1, 2]. The following presentation and device classification is to a great extent based on these works.

In Fig. 7.1, a selection of modern medical devices is shown. In the top row, we see a glucose monitor, a digital thermometer, and digital scales. In the second row, there is a fitness tracker and a heart monitor showing an electrocardiogram. In the bottom row, active high-risk, high-safety devices are depicted: an implantable defibrillator and a glucose management system.

Today's medicine is based on treating already present diseases and on the face-to-face contact with the doctor. Introducing smart medical devices will pave the road toward telemedicine and data-driven medicine. This will cause a paradigm change in the organization of our healthcare system [3, 4]. In the Internet age, one of the directions is toward predictive, preventive, personalized, and participatory (P4) medicine [5].

As medical devices change frequently and operate under varying conditions, evaluating their efficacy and safety is a challenge [6]. Comprehensively collecting and analyzing data would help, although it is a delicate task regarding privacy protection. A registry for cardiovascular device evaluation is presented in [7]. A framework for secure data exchange between many cooperating parties for this purpose is presented in [8]. Ethical aspects are discussed in [9].

Finding knowledge in data is a challenging task. It is easy to take a (spurious) correlation for causation. A harsh reminder about the pitfalls of statistical analysis can be found in [10]. Many studies are based on small samples or are biased, by financial interests and prejudices. It is easier to publish novel and spectacular results than simply true ones.

The problem of data quality is analyzed profoundly in [11]. In the case of medical data, the best data are obtained from randomized control trials (RCT), with large samples, good randomizing, and blinding. Also the provenance is important: how reliable is the source, are the data not biased, is collecting of data carefully managed? There are efforts to automatize quality verification with quality assessment tools (QAT). The problem is that the tools themselves have to be verified. Many are now on the market, but the results vary from product to product, making their reliability disputable.

Technical aspects of managing sensor data are presented in [12]. The authors treat such subjects as acquisition and cleansing, query processing, detecting events in streams, mining sensor data streams, and real-time data analytics.

In order to better delineate the subject to be discussed in this text, we will present several classifications of smart medical devices and the contexts of their use. The devices contain:

- sensors and/or
- actuators

---

[1]www.weforum.org/events/world-economic-forum-annual-meeting-2017/sessions/future-of-medicine-2017, Accessed January 22, 2016.

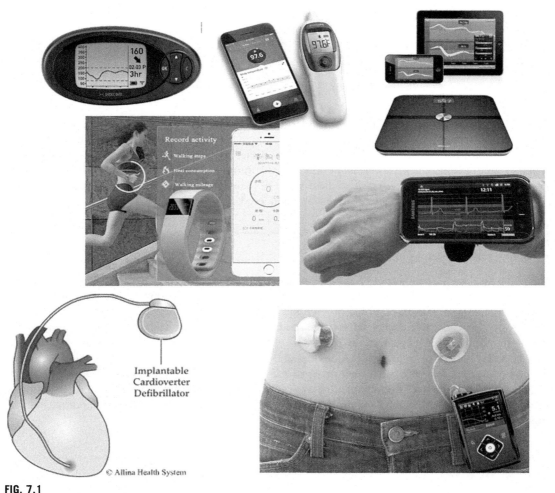

FIG. 7.1

Modern medical devices.

The sensors measure vital signals, and the actuators act directly on the human body, like pacemakers or insulin pumps. The data from the sensors result in an action on the body, but in wider loops, for example, influencing the decisions of the doctor or of the user/patient.

As for the placement, a device can be:

- related to the body (wearable/implantable)
- related to the location (ambient assisted living [AAL])

Body-related devices are often connected in a wireless body area network (BAN), with a data aggregator (smartphone) providing the connection with external partners. AAL applications are typically analyzed as ensuring secure home stay of an individual frail, elderly person, but actually they detect the activities of anybody present in the location.

A further classification relates to the safety category:

- lifestyle, fitness
- periodic measurements, health monitoring
- life saving

These classes correspond to the approval requirements of the concerned authorities (like the Food and Drug Administration [FDA] in the United States). The first class is freely available on the market, having to meet only the basic safety conditions of a consumer product. For other classes a registration is required up to formal certifications and clinical trials. In order to avoid costly and difficult approval procedures, some producers *suggest* medical value of their products but do not declare it formally.

Moreover, the following factors should be considered:

- severity of the disease
- necessity of the measurements/actions
- timing requirements

We will treat a device that helps to fight obesity and a life-saving device for a patient with an acute risk of a heart attack differently. We will use a different approach, starting with the composition of the design team and the organization of the project. Depending on the severity of the treated case, the collected values can be optional or absolutely required. In some cases, like heart monitoring, we face strict timing requirements. We speak of the *golden hour*, after which the help may be too late and useless.

The development of smart medical devices and their applications is very rapid, multidirectional, and quite chaotic. The border between personal fitness trackers and serious medical devices is especially blurred. If a device is an element of a medical therapy or its measurements are used to make medical decisions, their quality, reliability, and safety has to be monitored. Several important-related issues are treated in the rest of the chapter.

7.3 RISK ANALYSIS

In the discussion of security we want to take a broad view on the possible risks to the patient connected with the use of smart medical devices. In order to focus our attention we will list several sources of risk that we will analyze later more in detail:

- malicious action
- neglect
- external factors

We especially want to stress the importance of the neglect of developers that leads to bad design and hidden failures. A catastrophic effect can be caused by a trivial error. A classic example is the disaster of the Space Shuttle Challenger on January 28, 1986, that cost the lives of seven people.[2] It was caused

---

[2]en.wikipedia.org/wiki/Space_Shuttle_Challenger_disaster, Accessed January 22, 2017.

by rubber O-rings that should seal the shuttle and that lost elasticity at temperatures below the freezing point. The flow of information in this example is very instructive: the engineers knew about the problem, the management was aware but did not postpone the flight, and on the critical night the temperature was lower than ever before.

In risk analysis two factors play a role:

- probability of failure
- consequences of failure

The lesson from the Challenger disaster is that if we design a device or system that should save health and life or could damage it, we have to adapt the entire design process to meet the safety requirements. Another lesson is the influence of the changing environment. Even though the design flaw had existed all the time, the astronauts had been lucky because the lowest temperature at launch until that time had been 12°C.

In the following sections, we will take a closer look at specific security and privacy risks. Here we want to stress the importance of the skills of the people who develop and operate novel medical devices. First, as shown in Fig. 7.2, developing high-risk/high-reliability devices requires a specific mindset, focused on safety issues. A general software engineering course concentrates mostly on the development itself, which, in the face of such rapid changes in the tools used, is complex enough. This rapid progress makes it difficult to master the tools in detail. Thus a more conservative approach is advisable: use only well-known tools and spend more time on testing. This goes against the common practice to stay always at the front.

A similar goal conflict faces management: quality and reliability have to be guaranteed in the pursuit to lead the market.

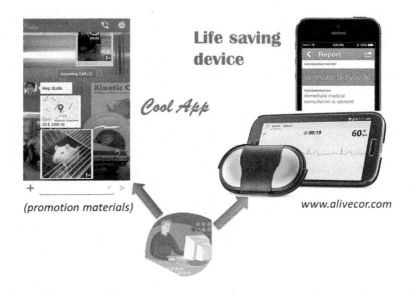

FIG. 7.2

Different mindsets.

Moreover, the development of medical devices and their successful introduction into the market requires a good understanding of many other areas:

- sensor/actuator design
- biochemistry
- medical knowledge
  - diagnosis
  - health monitoring
  - emergency services
  - hospital workflow
  - evidence-based medicine
  - healthcare system organization
  - reimbursement
- IT skills
  - smartphone apps
  - real-time processing
  - heterogeneous networks
  - security and privacy
  - Big-Data analysis
  - machine learning
- FDA approval process
- organization of clinical trials
- legal aspects
- business plan

Evidently, such a wide spectrum of skills is not found in a single person. It has to be, however, present either in the team, or acquired externally.

A good introduction to risk evaluation and how it leads to good decisions is given in [13]. The author shows how to discriminate real risks from the apparent ones. He warns how to detect manipulating conditional probabilities, using the example of breast screening. He shows a deliberate misuse of statistics by vested interests, notably the pharmaceutical industry, who cherry-pick their sample groups to "prove" the efficacy of their products.

A useful view for healthcare professionals on the usefulness and safety of the new technologies can be found in [14]. Telemedicine increases access to medical services for underserved populations, for example, in rural areas. It improves safety in critical care medicine and reduces medication errors. However, many application developers have no medical training and the applications do not undergo rigorous evaluation.

American (FDA) regulation of medical devices and adverse event reporting is presented in [15]. A comparison between the American and European safety regulations for medical devices is given in [16], with respect to mandate, centralization, data requirements, transparency, funding, and access. The book [17] gives a detailed picture of regulatory compliance and data security and privacy of electronic medical records.

In [18], an overview is given of the methodology of analyzing safety-critical computer failures of medical devices. The authors investigate the causes of failures in computer-based medical devices and their impact on patients, by analyzing human-written descriptions of recalls and adverse event reports, obtained from public FDA databases. They characterize failures by deriving fault classes, failure modes, and recovery actions. Their analysis identifies safety issues in life-critical medical devices and provides insights on the design challenges.

A good example is seen in [19] of risk analysis of hemodialysis devices, performed by a multidisciplinary team involving engineers and clinical experts. The authors identified a potential harm list. They categorized the hazardous situations from negligible to catastrophic and correlated them to the deviation of a specific device parameter. Their multidisciplinary evaluation for severity of hazards (MESH) method contributes to a better link between engineering and clinical perspectives.

7.4 CYBER-PHYSICAL-SOCIAL SYSTEMS

Smart medical devices based on health sensors rarely can be considered in isolation. Even simple sensors that just measure a parameter and display it to the user may be more or less usable, depending on the design of their user interface and on the skills and the habits of the user, his/her age, or medical status. The problem of the environment is even sharper when the device interacts extensively with this environment. Therefore we speak about cyber-physical-social systems [20], that have an IT part (computing and communications), but also have physical properties like any other object and are operated by humans with unpredictable behavior.

In real-life operation, a device (or a set of devices) is embedded in an environment consisting of nature, other technical systems, and humans—some of them friendly, other ones malicious (Fig. 7.3).

If a heart monitoring device cooperates with the emergency systems (Fig. 7.4), the effect depends on seamless transfer of data, predefined cooperation between hospitals, cognitive skills and

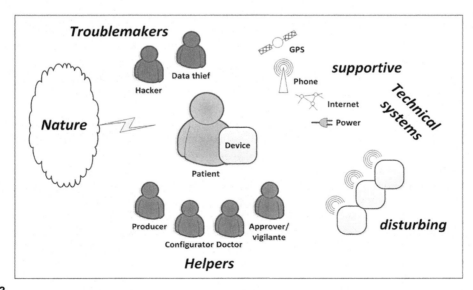

FIG. 7.3

Medical device and its environment.

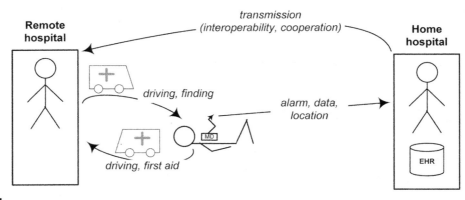

FIG. 7.4

Emergency services.

information state of the operators, and finally on traffic jams. In the case of a heart attack, the final result is positive if the whole process is concluded during *the golden hour*.

7.5 MACHINE ETHICS
7.5.1 ETHICAL DILEMMAS

It may not be immediately obvious, but the machines we build make decisions, and every decision regarding humans has ethical aspects. This subject, machine ethics, finds a broad coverage in [21]. The authors argue that as robots take on more and more responsibility, they must be programmed with moral decision-making abilities. They examine the challenge of building artificial moral agents that extend human decision making and ethics.

Even if the machine is just an insulin pump, the applied algorithm decides about the dosing (bolus); its quantity, shape, and timing. The problem becomes complicated if the process does not run as expected and we have to handle special conditions.

Let us consider a heart monitoring system that sends alarms to the hospital in an emergency. The electrocardiogram is analyzed continuously; an abnormal shape or frequency is a warning signal. The system can call an ambulance, giving the physical location, and the medical status of the patient. We should not miss the real danger but we cannot send ambulances without necessity. Even 10% of false positives would be damaging to the reputation of the system. We can improve the selectivity of the system by including more sensors and accept only a combination of signals.

Sending alarms by this system depends on the presence of the phone signal with the data roaming option active. What should we do if this signal is missing? On the one hand, the patient leaves the safety zone and we should warn him/her; on the other hand, issuing too many warnings will make him/her even more nervous and will be damaging to his/her health.

Let us take another example: a system distributing medicine to the patient with the possibility of verifying if the pill is actually taken and swallowed. This works perfectly if the patient complies with the recommendations. But what should we do—or rather what should the machine built by us do—when the patient does not want to take his/her pills? The patient may have various reasons. The patient may not be fully conscious, or may have dementia and require external guidance. He/she may be perfectly conscious and be suffering from strong side effects of this medicine. The medicine could be absolutely necessary, where not taking it would cause an immediate relapse, or it may be just a useful supplement, but basically optional. What should be done; what is the good decision? The system may:

- accept the patient's decision
- force him/her to take the pill
- give him/her an electroshock
- inform the doctor
- inform the family

We can see the following basic dilemmas:

- patient dignity vs. rules and regulations
- patient safety vs. patient freedom

Who has to decide? Effectively it may be:

- medical/ethical council
- patients' organization
- public debate
- producer:
  - system analyst
  - project manager
  - (junior) programmer $\Rightarrow$ it "just happens"

The answer is not evident. It may depend on the local culture. In the West, typically the decision of the individual is considered more important, but in the East family and community play an important role.

Even if we want to respect the will of the individual, this priority is not absolute if this will is damaging to his/her health. Especially in the case of the elderly suffering from dementia, often the decisions are made by others for the (apparent) good of the patients [22]. However, people on the dominant, healthy side may tend to treat the patient like an object. What if stopping smoking is for the patient more painful than the risk of shorter life?

We have to be aware that the new sensing technologies, both wearable/implantable and placed in the home (ambient assisted living), connected online to the outside world, are a massive invasion into the private sphere. External observers may have a continuous view of the vital parameters of the patient and the chemistry of his/her body. They know if and when he/she takes a shower, goes outside or receives visits. If it appears that such a vigilant level is necessary to care for the patient's health and life, those observers should act with full respect for the dignity of the patient and be very prudent when sharing sensitive information.

7.5.2 IMPLEMENTING ETHICS

When we agree upon our ethical principles, we have to implement them. Finally, in software an ethical decision can be reduced to a statement such as:

```
if (condition)
then {action1}
else {action2}
```

The problem is the precise definition of the conditions and the actions. To begin with, the programmer has to be aware of the consequences of his/her code. We stress it here because many applications, like smartphone apps, are *lightweight*—their errors or failures do not cause significant damage. This is also true for fitness monitoring. If the indicated step count is not exact and is different on different devices, it is not a problem. The situation is entirely different for applications having a direct influence on human life. Fig. 7.5 shows some examples of such situations. Therefore it is extremely important to well specify the conditions that trigger significant actions. For example, we have to be very careful when we decide to deliver an electric shock using a defibrillator. We should miss no real event but do it only when really necessary. This may require combining coordinated signals from many sensors. This condition may also be patient-dependent: an ECG shape that indicates an imminent danger for one patient might still be in the acceptable range for another.

If we adapt our systems using machine learning, we have to be aware that we give the ultimate control to the machine. This still may produce an algorithm better than we can achieve manually, but we depend on the learning procedure and on the training data. For example, if the patient's

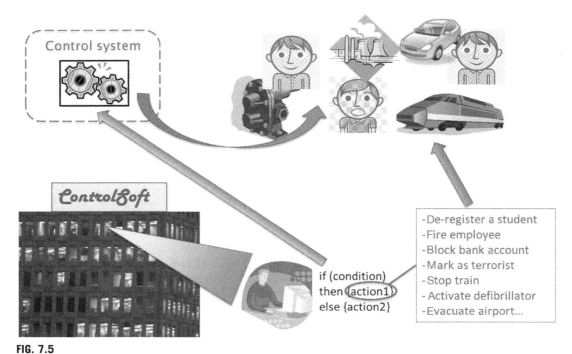

FIG. 7.5

Algorithm acting on the real world.

condition changes radically, what we can know and recognize immediately, the algorithm will need time to relearn and adapt to.

A special problem is the decisions that *come from nowhere*. We mean here the `if-then-else` blocks of code where the implications of the decision are not noticed by the programmer. Many systems are underspecified, and in the coding process the programmer meets some locations where a decision has to be made. For example, we may develop a system communicating with the external world, and define how to treat the data but never consider the case of a connection loss during the transmission. The system can interrupt the transfer, wait (with or without timeout) or beep to alarm the user. If the function is vital to the patient, the choice of the action is equally vital. Unfortunately, the programmer is often left alone, especially if he/she is working under deadline pressure. He/she will probably make a spontaneous decision, write some simple code (like logging the exceptions without proper handling) and the users may be surprised by the result.

A profound domain knowledge will help the programmer to make better decisions or at least to recognize the problem and ask competent people. This shows that in software engineering education an understanding of the real world should be taught, in addition to IT skills.

7.5.3 RESPONSIBILITY

For simple devices there are laws regulating product liability. Nowadays the problem is getting more complex. Cyber-physical systems depend on changing environments and the skills of unknown people, and also assume the existence and functioning of external systems. Learning systems develop their control algorithm based on data. After some time passes, the algorithm is unknown to the developers. Who is responsible for the actions of autonomous machines, if anyone at all, is an open question [23]. This problem becomes more visible as self-driving cars are ready for deployment. In the medical arena we face similar questions.

The problem of responsibility does not only concern physical action but also propagation of information. For an individual, the most spectacular cases concern commercially available DNA tests. A negative result may lead to doubtful decisions, like the preventive double mastectomy and removal of ovaries by Angelina Jolie, diagnosed with a rare mutation of the BRCA1 gene. The company 23andMe[3] had a long dialog with the FDA regarding approval of their tests. Among other things, they discussed what syndromes have to be detected and how the tests have to be promoted. It is irresponsible to suggest miraculous properties of a test and to give in a report the cancer risk or life expectancy without any comment or clear indication of the reliability of the result. This may destroy the life of the tested person. The rise and fall of the Theranos company[4] can serve as a warning to overambitious entrepreneurs and credulous customers. The company's innovative blood-testing technology made the CEO, Elizabeth Holmes, America's youngest female billionaire. However, their technology was so secret that no details were published or openly discussed. Finally it became apparent that even the test results were fake and the test could rather put lives in danger than be of help.

For the community, the facts, rules, and recommendations derived from the data will influence decisions of others. Therefore even in the scientific community there is a drive to publish spectacular

---

[3] www.23andme.com, Accessed January 22, 2017.
[4] www.zerohedge.com/news/2016-05-18/elizabeth-holmes-admits-theranos-technolgy-fraud-restates-voids-years-test-results, Accessed January 22, 2017.

results rather than simply true ones. So many apparent correlations are spurious. However, once published, they circulate without any further verification. If we come back to the case of Angelina Jolie, even if the individual decision was correct, the induced decisions made by other people are not necessarily so [24].

7.6 PHYSICAL SAFETY

Before we analyze the subtler aspects of security, we have to look at the device as a physical object. There are several physical factors that affect the safety of the patients. If the device has contact with the skin, especially for a long time, it is important that the material does not cause inflammation. For an implantable the risk is greater, because the contact is constant and with time tissue grows around the device, which makes replacement difficult. Some devices, like pacemakers, have thin probes that can break.

The surface of a sensor can become dirty, which will distort the measurement. If the sensor is placed on skin, the contact will be bad if the skin is too wet or too dry. If the device is manually placed by the patient, the correct placement will depend on his/her skills.

The devices that act on the body pose similar problems. An electrode that activates an organ has to be safely guided to the correct location and the contact has to be clean. If a device injects a substance into the body, its nozzle must not clog.

A wearable device can be effectively used by the patient, as long as the usage is comfortable enough. We speak here of patient's compliance. The patient will tend not to wear the device if it is too heavy or obtrusive. In the case of a stigmatizing disease, the device should not be visible to other people.

A device may also not be used properly if it requires too much attention and service. A simple example is ensuring that the batteries are not discharged. A smartphone needs to be charged at least daily, and we know that this can be a problem, especially on travel. If a patient has a set of several devices with different voltages, dimensions, and lifetimes, this task is a challenging one. The devices should at least alarm the patient about a low battery state. On travel, the patient must carry the complete collection of chargers, with plug adapters if necessary. Therefore the construction of low-power devices is very important, as well as energy harvesting or contactless energy transfer to an implanted device.

All those issues may seem trivial, but trivial issues like a broken electrode or empty battery can make even the best device useless.

7.7 SOFTWARE QUALITY

We have a relatively long experience with developing and using simple medical devices. The approval process and the failure reporting procedures are well defined. For example, if a device is significantly similar to an already approved device, the process is easier. However, if we add complex software to a known device, we introduce many new risk factors [25].

Problems may arise on any level. Algorithms can be wrong or inexact. In the decision process the situations may be incorrectly classified, and wrong actions may be taken. Adaptive/learning systems may show unpredictable behavior. Finally, complex, multitasking software may have implementation errors, especially if it was developed under time pressure (as it always is).

We have mentioned previously the necessity of an interdisciplinary approach and a safety-oriented attitude. This calls for a well-controlled project organization. Extensive testing is necessary, under changing conditions, with various users. Young, healthy developers must think of elderly, ill patients, having poor vision, tremor, or cognitive impairment. We should not forget the *black swans*, situations that should "never occur."

Flexibility and variability of the product and its environment are major issues. Big players in the market deliver monolithic solutions, fully controlled and supported by them. Smaller companies produce devices that can be combined with others into workable systems. Open systems permit one to select the best available elements and make them work together. The downside is that the responsibility for the overall function becomes unclear. It also has to be defined who actually configures the final system. A hospital has no such capacities. This shows again that introducing smart medical technologies can cause a disruptive change to the organization and business model of the healthcare system.

Innovative systems are improved frequently. Software on the devices will be updated, to extend the functions, fix the bugs, or patch the security holes. Therefore ensuring interoperability of protocols and data formats is a task in itself. Moreover, remote software update is a possible attack vector: instead of a valid patch malicious software can be sent. For this reason, after sale IT support is also necessary.

7.8 **IT SECURITY**

A good overview of the security issues of smart medical devices can be found in [26–28]. Those problems are known more generally in the IOT—small, smart, networked devices deployed in the environment. As this field is new, the pace of progress is immense and many inexperienced startups are active, and common practices known in more established areas of IT are not well followed. The "2017 Study on Mobile and IoT Application Security," conducted by the Ponemon Institute and sponsored by IBM Security and Arxan Technologies states that 80% of IoT applications are not tested for security flaws.[5] Bruce Schneier, a renowned security expert, says that "The Internet of Things is wildly insecure—and often unpatchable."[6]

There are several reasons for this state. The devices are typically sold as small and cheap hardware devices, but actually they are complex computing appliances. Their software needs to be supported and patched. If they are not updated for new security flaws, their security erodes with time. Equally, if they cooperate with other devices from other producers and data exchange protocols or data formats are updated, the interoperability will be broken. We see that time and change play major roles.

The security problems arise with the local storage of data, data transmission, and storage on a central server.

Portable mobile devices may be physically destroyed, lost, or stolen. In this case personal data may be accessed if local protection has not been properly designed.

---

[5]media.scmagazine.com/documents/282/2017_study_mobile_and_iot_70394.pdf, Accessed January 22, 2017.
[6]www.schneier.com/essay-468.html, Accessed January 22, 2017.

Many networked medical devices still transfer data in plain text, which permits eavesdropping on personal data or activating a pacemaker from outside. Therefore encryption is a crucial requirement, which may be problematic for low-power devices with limited computing capacity. Moreover, especially in more flexible network configurations, ensuring correct management of security keys is difficult. We face here a trade-off between security and availability, If another device is not recognized as a valid partner, it may be a malicious attacker and the communication should be blocked. On the other hand, if the communication is blocked due to unknown or expired keys, but connection is necessary to ensure the life-saving function of the combined device, "tolerance" would be preferred. The decision is not easy.

Until now we spoke of the devices themselves and their own supporting systems. If the urban space is filled with more networked devices, we can face the problems of interaction, from electromagnetic noise via competing for bandwidth to wrongly addressed data packets.

7.9 PRIVACY

The storage and processing of health-related data is strictly regulated by law. In Europe, it is the General Data Protection Regulation (GDPR) (Regulation (EU) 2016/679) and in the United States the Health Insurance Portability and Accountability Act (HIPAA). The European regulation prohibits (Article 9.1) the processing of data revealing *race or ethnic origin, political opinions, religion or philosophical beliefs, sexual orientation or gender identity, trade-union membership and activities, and the processing of genetic or biometric data or data concerning health or sex life, administrative sanctions, judgments, criminal or suspected offences, convictions, or related security measures*. On the other hand, Article 81 gives a list of exceptions for processing health data, among others:

- using personal data, if necessary, preferably anonymously
  - preventive medicine
  - management of healthcare services
  - epidemics control
  - quality assurance
  - ensuring the quality and cost-effectiveness (evidence-based medicine)
- using anonymous data, with consent
  - public health
  - scientific research

The data collected and their purpose have to be well defined. Data have to be processed on systems developed according to the principle of *privacy by design*. The physical location of those systems and the institutions that operate them, including intermediaries, has to follow rules set by law. The users, depending on the access level, have to be known and registered, and clearly bound by professional secrecy. The user actions have to be trackable.

We see that, although the law principally stresses the protection of health data, nevertheless it respects their reuse for the public good, under certain conditions. As the data needs of the acceptable goals overlap, this leaves a space for a discussion. Several approaches and ways of thinking can here be observed:

- legal: protection of personal data
- governmental: safety control
- evidence-based medicine: effectiveness assessment
- scientific: knowledge from data
- industrial: quality control
- open data: data mining for all
- community: patients like me, sharing experience
- "Facebook": sharing with friends/anybody
- commercial: data as product

Those approaches are based on different value systems and are represented by different communities that pursue different goals. For some, technical progress and the free market are most important, and for others, patients' health or his/her privacy are paramount. All those approaches can to a certain degree be justified. They are partly contradictory and they call for a consensus and technical means to support a workable solution.

Those approaches vary mostly in the trade-off between hiding and disclosing data. There is no easy solution. For example, it is important to protect the patient's privacy but it is in his/her interest to gain medical knowledge from available data, also his/her data, although everybody would prefer only other people's data to be reused. This is not reasonable. Choosing not to share my data for research makes my case less well studied. In the case of precision medicine based on genetic markers, the size of the sample of people having the same combination of markers as me is much smaller than the global patients' population. Therefore not sharing my data makes the verification of my therapy method less reliable. The communities of patients are growing. They are aware that their common enemy is the disease and all medical progress is good for them all.

Evidently, *decent* ways of reusing data for public good all assume anonymizing of data. On the other hand, ingenious data miners can break this anonymity, especially when combining it with other sources of public Big Data. They may act with malicious intentions or just to prove their technical prowess. This problem is getting worse, as many biological features, like electrocardiogram (ECG), electroencephalogram (EEG) or even gait pattern are used to identify persons. The genetic information, necessary for medical reasoning, is itself a personal identifier.

Fig. 7.6 identifies the main stakeholders in the exchange of medical data. The trade-off between their legitimate interests is discussed in [29] and a more general analysis for handling personal data in an IoT environment is presented in [30].

The books [31, 32] provide a profound analysis of the trade-offs between privacy protection and the public good. The first book claims that what people really care about when they complain and protest that privacy has been violated is not the act of sharing information itself—most people understand that this is crucial to social life—but the inappropriate, improper sharing of information. The second book tries to answer questions like: What are the ethical and legal requirements for scientists and government officials seeking to use big data to serve the public good without harming individual citizens? What are the rules of engagement with these new data sources? What are the best ways to provide access while protecting confidentiality? Are there reasonable mechanisms to compensate citizens for privacy loss? In this discussion it covers many specific areas of social life.

The paper [29] discusses the specific case of reuse of medical data. In order to assess the quality and efficacy of the treatment, it is necessary to gather a possibly comprehensive collection of the treated patients. For some diseases, additional parameters, like demographics, are necessary. Even if the names

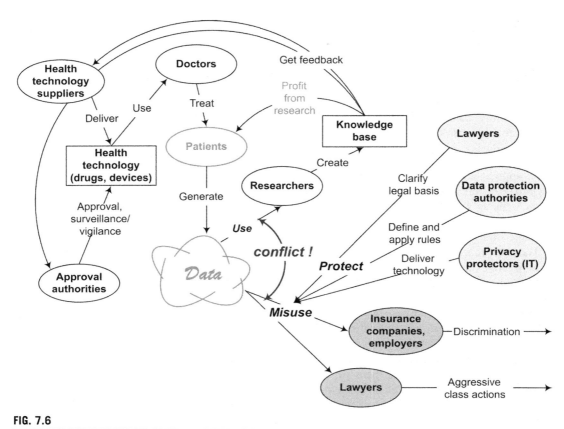

FIG. 7.6

Stakeholders in the use and reuse of medical data.

are removed, there is still a possibility of reidentification by a clever data miner. This is in direct conflict with the privacy protection rules. Technical means have to be found in order to ensure good quality of research results and still comply with the rules.

Reused data have to be anonymized, but Raghunathan [33] shows that this task goes well beyond just removing names. The paper [34] presents how to use the ECG for biometrics, that is, how to reidentify the patients on the basis of a medical parameter. Generally, the progress of data mining makes the efforts to anonymize patients difficult. Especially in genetic research (like precision medicine), the DNA fingerprint, basically containing medical information, not only permits a person to be identified, but it *is* the identifier itself.

7.10 RISK OF TECHNOLOGY MISUSE

Health data circulating in collection and analysis systems can be used and misused for various purposes. In particular, the usage of data in commercial systems is not well defined, and even if it is defined in some small-print documents, it is ignored and neglected by users. With a step-counting accelerometer built into my smartphone, am I aware who sees my data? Anonymous or personalized, aggregated or detailed? I can verify my step count and see my walk on the map. I can compare my values to my gender and age group. This means that those data are stored on a central server and can be used to analyze the public health or to observe me. Most users accustomed to the "Facebook" approach, used to publishing private data on the Net, do not care. In most cases this approach causes no problems. However, if the user makes an agreement with their insurance company to have a healthy lifestyle in exchange for a lower premium, the numbers obtain a monetary value. Many would complain of being observed, but the ethical case is not so clear. Lowering the premium, or cheating on the promised lifestyle: neither is correct. We must, however, be aware that, depending on the granularity of the data transferred, the user's location at any time may be disclosed. At any rate, it is stored on the server, and even if the operator wants to protect it, this information can be stolen. If we consider catastrophic options, we should take into account the possibility of a future totalitarian government that uses the already deployed technology for comprehensive tracking of its citizens.

Let us analyze a similar case. Let us assume that we can monitor the chemical composition of blood and detect specific substances. It may be nicotine, and the user can make an agreement with the insurance company, as in the previous example. The difference is that smoking may be considered an asocial activity, and what started as a voluntary option may with time become obligatory. Moreover, having nicotine traces in the blood could be punished. This leads to an ethical dilemma between the right to organize personal life at will and the benefit of the society defined by a controlled or uncontrolled authority.

Moreover, this technology can be transferred to a country with different values. We can accept mild enforcing of nonsmoking, but would have problems with detecting alcohol consumption, consumption if possibly with death. A similar problem has arisen with prenatal tests initially intended to be used for detection of severe genetic disease, but being used in other cultures to favor male offspring.

This shows that once the technology is there, it is difficult to predict and control its use [35].

7.11 ENHANCED HUMANS

As a possible consequence of the massive development of wearable and implantable devices, enhanced humans (or transhumans) may arise. Actually, if we take a historical perspective, we are technologically enhanced humans already. An obvious example of such enhancement would be eyeglasses. But only our homes (ambient assisted living) and wearable clothes allow us to live outside the African savanna. We simply do not notice these things anymore. As for the newer technology, our smartphones permit us to consult Wikipedia or timetables when walking, and we become accustomed to it, and then addicted and dependent.

(Remark: this passage corresponds to the technology state of 2018. In 10 years will only be of historical value.)

In a similar way, wearable and implantable devices will permit constant monitoring of our vital functions and ways to enhance them. At the upper end, it will be easier to be active despite having a chronic disease. A system monitoring the health function will warn the patient or call for help in an emergency. Deep brain stimulation will make life bearable to patients with neurodegenerative diseases. We can continue this list at will; the age of smart medicine is only beginning.

The price is the exposure to new risks, like attacks on a medical device by a hacker. In truth, every helping technology makes us dependent, but without this technology severe medical conditions could impair our lives or cause death.

7.12 CONCLUSIONS AND FUTURE WORK

The complexity and diversity of the area of smart medical devices, together with the rapid pace of development, make it difficult if not virtually impossible for an individual to grasp it entirely and profoundly. In this chapter we tried to raise the reader's awareness of related issues.

As every developer, operator, or supporter has their own specific view, it is useful to be open on other perspectives. It is a grave error to concentrate on our own skills and neglect other aspects. For example, a smart device project starts with a novel sensor concept and the initial team consists of sensorics specialists. If they intend to convert the idea into a usable product, they have to involve specialists with expertise in other fields and a leader with a broad view on all important aspects of the development, engineering, approval, and marketing process. A device communicating over a wireless link can be hacked if not adequately protected. It would not be surprising if this issue were not known at all to the sensorics specialists. Equally, they probably do not know the hospital procedures, and this knowledge is necessary to make a product that is useful in an actual therapy and that will be applied by doctors.

The goal of developing and effectively using smart medical devices is a moving target. Enclosing them in a system that is secure as a whole, protects the privacy of the patients, and resolves the possible ethical issues in a controlled way in agreement with well-defined ethical rules is a highly interdisciplinary task. Developing a commercial product in the medical field goes well beyond gathering funds and finding investors. All persons involved have to maintain an interdisciplinary approach, flexibility, and an open mind.

Current university education concentrates on providing training in relatively narrow specialties. Adding basics in other natural sciences is certainly useful. It is, however, even more important to teach collaboration on an interdisciplinary team, the ability to listen to specialists from widely different areas and to make one's own ideas understandable to them. Developing and implementing such programs is a major direction for future work.

REFERENCES

[1] Maharatna K, Bonfiglio S. Systems design for remote healthcare. New York: Springer Science & Business Media; 2013.

[2] Cavallari R, Martelli F, Rosini R, Buratti C, Verdone R. A survey on wireless body area networks: technologies and design challenges. IEEE Commun Surv Tutor 2014;99:1–23. https://doi.org/10.1109/SURV.2014.012214.00007.

[3] Wachter R. The digital doctor. New York: McGraw-Hill Education; 2015.

[4] Vallancien G. La médecine sans médecin? Le numérique au service du malade. Paris: Editions Gallimard; 2015.

[5] Hood L. Systems biology and P4 medicine: past, present, and future. Rambam Maimonides Med J 2013;4(2). https://doi.org/10.5041/RMMJ.10112.

[6] Becker KM, Whyte JJ. Clinical evaluation of medical devices: principles and case studies. New York: Springer Science & Business Media; 2007.

[7] Sedrakyan A, Marinac-Dabic D, Holmes DR. The international registry infrastructure for cardiovascular device evaluation and surveillance. JAMA 2013;310(3):257–9.

[8] Sliwa J, Benoist E. Research and engineering roadmap for development and deployment of smart medical devices. In: HEALTHINF; 2015.

[9] Meulen RHJ, ter Meulen R, Biller-Andorno N, Lenk C, Lie R. Evidence-based practice in medicine and health care: a discussion of the ethical issues. New York: Springer; 2005.

[10] Ioannidis JPA. Why most published research findings are false. Chance 2005;18(4):40–7.

[11] Floridi L, Illari P. The philosophy of information quality. vol. 358. New York: Springer; 2014.

[12] Aggarwal CC. Managing and mining sensor data. New York: Springer Science & Business Media; 2013.

[13] Gigerenzer G. Risk savvy: how to make good decisions. London: Penguin; 2014.

[14] Agboola SO, Bates DW, Kvedar JC. Digital health and patient safety. JAMA 2016;315(16):1697–8.

[15] Teow N, Siegel SJ. FDA regulation of medical devices and medical device reporting. Pharmaceut Reg Affairs 2013;2(110):2.

[16] Kesselheim AS, Rajan PV, et al. Regulating incremental innovation in medical devices. BMJ 2014;349: g5303.

[17] Robichau BP. Healthcare information privacy and security: regulatory compliance and data security in the age of electronic health records. New York: Apress; 2014.

[18] Alemzadeh H, Iyer RK, Kalbarczyk Z, Raman J. Analysis of safety-critical computer failures in medical devices. IEEE Security Privacy 2013;11(4):14–26.

[19] Lodi CA, Vasta A, Hegbrant MA, Bosch JP, Paolini F, Garzotto F, et al. Multidisciplinary evaluation for severity of hazards applied to hemodialysis devices: an original risk analysis method. Clin J Am Soc Nephrol 2010;5: CJN-01740210.

[20] Suh SC, Tanik UJ, Carbone JN, Eroglu A. Applied cyber-physical systems. New York: Springer Science & Business Media; 2013.

[21] Wallach W, Allen C. Moral machines: teaching robots right from wrong. Oxford: Oxford University Press; 2008.

[22] Jones CM. Preserving life, destroying privacy: PICT and the elderly. In: Emerging pervasive information and communication technologies (PICT). Springer; 2014. p. 89–99.

[23] Miller KW. Applying "moral responsibility" for computing artifacts to PICT. In: Emerging pervasive information and communication technologies (PICT). Springer; 2014. p. 193–207.

[24] Evans DG, Barwell J, Eccles DM, Collins A, Izatt L, Jacobs C, et al. The Angelina Jolie effect: how high celebrity profile can have a major impact on provision of cancer related services. Breast Cancer Res 2014;16:442.

[25] Fu K. Trustworthy medical device software. In: Public health effectiveness of the FDA 510(k) clearance process: measuring postmarket performance and other select topics: workshop report, IOM (Institute of Medicine): National Academies Press; 2011.

[26] Fu K, Blum J. Controlling for cybersecurity risks of medical device software. Commun ACM 2013;56(10):35–7. https://doi.org/10.1145/2508701.

[27] La Polla M, Martinelli F, Sgandurra D. A survey on security for mobile devices. IEEE Commun Surv Tutor 2013;15(1):446–71.

[28] Hei X, Du X. Security for wireless implantable medical devices. New York: Springer; 2013.

[29] Sliwa J, Benoist E. Medical evaluative research and privacy protection. In: Developments in e-systems engineering (DeSE); 2012.

[30] Sliwa J. A generalized framework for multi-party data exchange for IoT systems. In: 2016 30th international conference on advanced information networking and applications workshops (WAINA), IEEE; 2016. p. 193–8.

[31] Nissenbaum H. Privacy in context: technology, policy, and the integrity of social life. Stanford: Stanford University Press; 2009.

[32] Lane J, Stodden V, Bender S, Nissenbaum H. Privacy, big data, and the public good: frameworks for engagement. Cambridge: Cambridge University Press; 2014.

[33] Raghunathan B. The complete book of data anonymization: from planning to implementation. Boca Raton: CRC Press; 2013.

[34] Silva H, Lourenço A, Canento F, Fred ALN, Raposo N. ECG biometrics: principles and applications. In: Biosignals. p. 215–20.

[35] Sliwa J. Do we need a global brain? tripleC: Cogn Commun Coop 2012;11(1):107–16.

DIAGNOSING MEDICAL CONDITIONS USING RULE-BASED CLASSIFIERS

8

Juana Canul-Reich, Betania Hernández-Ocaña, José Hernández-Torruco
Universidad Juárez Autónoma de Tabasco, Cunduacan, Tabasco, Mexico

8.1 INTRODUCTION

Fast and accurate disease diagnosis is a critical task in medicine. This process has a significant impact on patients' recovery. The quicker a proper diagnosis is provided, the sooner physicians can take actions leading to patients well-being. Currently, machine learning techniques allow creating prediction models applicable in diagnosis decisions. These computational models have some advantages over traditional diagnosis methods; they are cheap, simple, automatic. Also they avoid some invasive tests. Some examples of machine learning techniques in disease diagnosis can be found in [1–3].

In this study, a comparison of methods for disease diagnosis was performed by using three rule-based classifiers (JRip, OneR, and PART) against kNN and SVM. Rule-based classifiers provide simple prediction models easily interpretable by humans. On the other hand, kNN and SVM are among the most widely used algorithms in pattern recognition and classification tasks.

We selected diabetes, arrhythmia, hepatitis, and Parkinson datasets for experiments. All datasets come from the UCI repository. The study aims to identify the best performance in machine learning-based prediction models as support systems to physicians in diagnosis tasks.

This chapter is organized as follows. In Section 8.2, we describe datasets used for experiments, followed by a brief description of metrics used in performance evaluation, and a brief description of the classifiers follows. Then, the experimental design and the parameter-tuning procedure of classification algorithms are provided. In Section 8.3, we show and discuss the experimental results. Finally, in Section 8.4, we summarize conclusions of the study and also suggest some future work.

8.2 DATASETS

The datasets we used to carry out this study included the following diseases: diabetes, hepatitis, Parkinson, and arrhythmia, obtained from the UCI Machine Learning repository [4]. No preprocessing steps were conducted on any dataset.

Table 8.1 Description of Datasets Used in Experiments

Disease Dataset	No. of Instances	No. of Attributes	No. of Classes
Diabetes	768	9	2
Hepatitis	155	20	2
Parkinson	197	23	2
Arrhythmia	452	279	16

The diabetes dataset is the so-called "Pima Indians Diabetes Database" donated by Vicente Sigillito of Johns Hopkins University in May 1990, whose source is the National Institute of Diabetes and Digestive and Kidney Diseases.

The hepatitis dataset was donated by Gail Gong of Carnegie-Mellon University, through the Jozef Stefan Institute in Slovenia, in November 1988.

The Parkinson dataset was created by Max Little, affiliated with the University of Oxford, in collaboration with the National Center for Voice and Speech, Denver, Colorado, who recorded the speech signals [5].

The arrhythmia dataset, called the cardiac arrhythmia database, has been donated by researchers from the University of Bilkent, in Ankara, Turkey. Details of each dataset are given in Table 8.1

8.3 METHODS

We used three rule-based algorithms to create diagnosis models using JRip, OneR, and PART. We also used the algorithms k-nearest neighbor (kNN) and support vector machine (SVM), which generate nonrule-based classification (diagnosis) models. Our interest was to compare the performance of rule-based diagnosis models against others.

On the other hand, for performance analysis, we used the following metrics: balanced accuracy, accuracy, sensitivity, specificity, and kappa statistic.

All these methods are briefly described in the following sections.

8.3.1 CLASSIFICATION METHODS

8.3.1.1 JRip

This is an implementation of the RIPPER algorithm in the Weka tool. RIPPER stands for Repeated Incremental Pruning to Produce Error Reduction. It was introduced by Cohen [6]. Classes are analyzed in order of increasing size. An initial set of rules is generated for the class using incremental reduced error. All examples in the training set belonging to the analyzing class are treated by generating a set of rules that, altogether, cover all these examples. Then JRip continues analyzing the next class and repeats the same procedure. It goes on until all classes have been covered [7, 8].

8.3.1.2 OneR

This method is called OneR for one rule. It consists of making rules testing a single attribute and branching down according to the possible values of the attribute. That is, each attribute value opens up a new branch. The class assigned by each branch will be the most frequently occurring class in the training data. Then the error rate for each rule can be determined, and the rule with minimum error rate is the best choice [8].

8.3.1.3 PART

PART (for partial decision tree) is another approach for rule induction. A pruned decision tree is built for the current set of instances. Then, the leaf with the largest coverage is converted into a rule, and the tree is discarded. Instances covered by this rule are removed. This process is recursively repeated for the remaining instances until none are left.

The key idea is to build partial decision trees instead of fully explored ones, since only one rule will be derived from each of them. A partial decision tree is a regular tree with undefined subtrees. That is, its construction process ends as soon as a subtree not simplifiable any further is found. Once the partial tree is obtained, a single rule is extracted from it. For a thorough description of this method see [8].

8.3.1.4 kNN

The kNN classifier assigns an unseen instance to the class with most instances among its kNNs. Generally, k takes an odd value to avoid ties. There are some distance metrics d to measure the distance between two instances with n attributes, such as Euclidean distance and Manhattan distance, among others.

The value given to k is important. If k is too low, we may be fitted to the possible noise in the data. If k is too high, the method may miss the local structure in the data. So, a good technique is to experiment with some k values, which involves using the training data to create a model and a separate validation set to look at its error rate. The k value chosen would be the one with the minimum error rate on the validation set. Typically, values of k lie in the range of 1–20 [9, 10].

8.3.1.5 SVM

SVMs were first introduced by Vapnik [11] in 1998, and initially they worked on binary datasets only. The SVM consists of mapping the training data, known as input space, into a higher dimensional space, known as feature space, by applying a kernel function; and finding a maximum margin hyperplane in the feature space that separates the two classes. Maximization of margin between the two classes results in improvement of classification performance. SVMs can typically use a polynomial, linear, radial basis function, and a sigmoid kernel.

A regularization parameter C value is chosen by the user. It is used to penalize errors. A small C value means errors are allowed and a larger margin is found by the SVM [12]. A thorough description of SVMs can be found in [13, 14].

8.3.2 PERFORMANCE MEASURES

8.3.2.1 Accuracy

This is the most typical performance metric used in classification. It is the ratio of correctly classified instances to the total number of instances in the dataset.

8.3.2.2 Balanced accuracy

This is a classification performance metric conveniently applied when imbalanced datasets are used in experiments. It is defined as follows:

$$\text{Balanced accuracy} = \left(\frac{\text{TP}}{\text{TP}+\text{FN}} + \frac{\text{TN}}{\text{FP}+\text{TN}} \right) \Big/ 2 \tag{8.1}$$

where TP is the true positive, FN is the false negative, TN is the true negative, and FP is the false positive.

8.3.2.3 Sensitivity

Sensitivity is a metric to indicate the goodness of a classifier to classify true positives. That is, in a diagnostic test, it would be the ability to accurately classify ill people. It is defined as follows:

$$\text{Sensitivity} = \frac{TP}{TP + FN} \qquad (8.2)$$

8.3.2.4 Specificity

Specificity is a metric to indicate the goodness of a classifier to identify true negatives. That is, in a diagnostic test, it would be the ability to accurately classify healthy people. It is defined as follows:

$$\text{Specificity} = \frac{TN}{TN + FP} \qquad (8.3)$$

8.3.2.5 Kappa statistic

This is a measure introduced by Cohen [15]. It measures the agreement between predicted versus ground truth classifications of a dataset. At the same time, it corrects randomly occurring agreement [8].

According to Cohen [16] the kappa statistic lies in the range from 0 to 1 as follows:

0 = agreement equivalent to chance
0.1–0.20 = slight agreement
0.21–0.40 = fair agreement
0.41–0.60 = moderate agreement
0.61–0.80 = substantial agreement
0.81–0.99 = near perfect agreement
1 = perfect agreement

8.3.2.6 10-Fold cross-validation

This evaluation technique partitions the original dataset into 10 mutually exclusive equal-sized splits. The whole process will take 10 iterations. At each iteration, 9 of 10 partitions are used for training, and 1 partition is chosen for testing so that each partition is used for testing exactly once. The error rate is computed for each iteration. The total error rate is the summation of errors across all iterations.

8.4 RESULTS

kNN requires the optimization of two parameters, which are the number of neighbors (k) and the value for distance (d). For this work, the Manhattan and Euclidean distances were used, in order to define with which of the two distance values the best results are obtained. For k, we took values close to the square root of the number of instances of the dataset in question; for example, for the hepatitis dataset with 155 instances, the square root being 12.45, the values of k analyzed were 5–20. The optimization results of kNN parameters for different values of k and d, for each disease, are shown in Fig. 8.1. Values for k and d with the highest balanced accuracy were used for further kNN classification experiments.

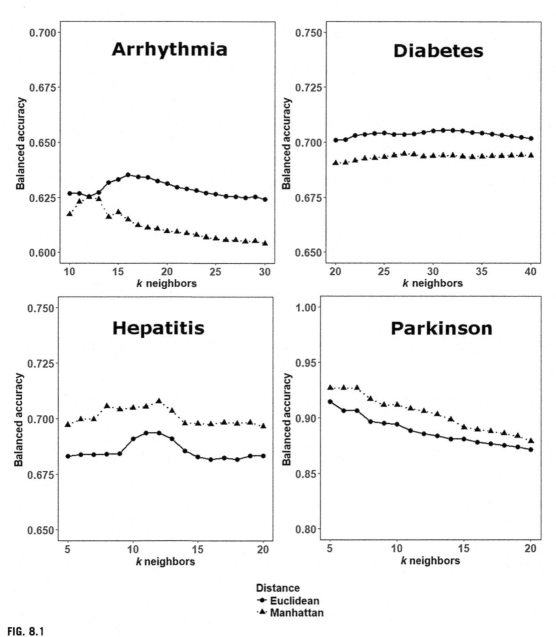

FIG. 8.1

kNN tuning for diabetes, arrhythmia, Parkinson, and hepatitis.

SVM optimization consisted of using increasing values of the variable C to obtain the highest value of balanced accuracy. The values used were 0.001, 0.01, 0.1, 1, 10, 50, 80, and 100. The results are shown in Table 8.2.

In Table 8.3, corresponding to the performance measures of the prognosis of hepatitis, it can be observed that the highest average obtained for the balanced accuracy was found to be slightly higher than 0.70. SVMLin had the highest accuracy value with 0.80. The specificity metric achieved the highest average of all performance metrics exceeding 0.90, while kappa obtained the lowest average of 0.0564. The standard deviation in the balanced accuracy of kNN shows that this classifier obtained similar results across the 30 runs. Two classifiers were above 0.70 in balanced accuracy, SVMLin and kNN, and neither of them are rule-based classifiers. The best rule-based classifier was JRip with a value of 0.6502 in balanced accuracy.

Table 8.2 SVMLin Tuning for Each Disease

C Values	Diabetes	Arrhythmia	Parkinson	Hepatitis
0.001	0.5000	0.6362	0.5000	0.5000
0.01	0.7154	0.6562	0.7260	0.5000
0.1	0.7207	**0.6863**	0.7660	**0.7232**
1	**0.7225**	0.6805	0.7721	0.6967
10	0.7221	0.6805	0.8101	0.6974
50	0.7221	0.6805	**0.8115**	0.6974
80	0.7220	0.6805	0.8091	0.6974
100	0.7220	0.6805	0.8089	0.6974

Note: *The highest balanced accuracy appears in bold.*

Table 8.3 Average Performance Measures After 30 Runs for Hepatitis

Classifier	Balanced Accuracy	Accuracy	Sensitivity	Specificity	Kappa
JRip	0.6502	0.8104	0.3833	0.9172	0.3363
(SD)	0.0319	0.0206	0.0604	0.0209	0.0694
kNN	0.7078	0.8378	0.4911	0.9244	0.4503
(SD)	0.0170	0.0096	0.0327	0.0087	0.0336
OneR	0.5427	0.5444	**0.5455**	0.5455	0.0564
(SD)	0.0209	0.0135	0.0185	0.0185	0.0273
PART	0.5570	0.7886	0.1711	**0.9430**	0.1453
(SD)	0.0260	0.0133	0.0604	0.0195	0.0630
SVMLin	**0.7232**	**0.8411**	0.5267	0.9197	**0.4731**
(SD)	0.0266	0.0154	0.0506	0.0138	0.0524

Note: *The highest values appear in bold.*
SD, *standard deviation.*

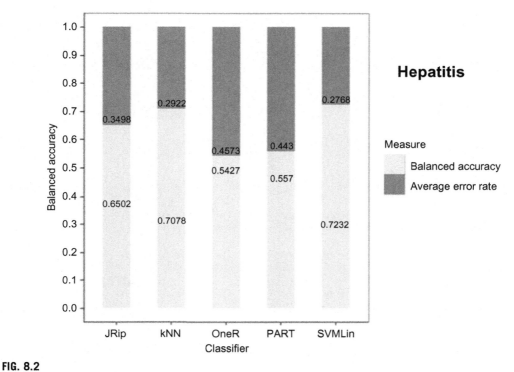

FIG. 8.2

Balanced accuracy and average error rate across 30 runs for hepatitis.

In Fig. 8.2 we show each of the classifiers with the balanced accuracy as well as the average error rate for hepatitis. The SVMLin classifier obtained the best result with a balanced accuracy of 0.7232, while for the rest of the classifiers the balanced accuracy varied between 0.5427 and 0.7078, with OneR being the classifier with the lowest average result. That is, SVMLin is the classifier that obtained the most hits with 72.32% in the test cases.

Fig. 8.3 shows the behavior of the balanced accuracy of each of the classifiers across 30 runs for hepatitis. The balanced accuracy of OneR ranged from 0.5083 to 0.5833. The balanced accuracy of JRip varied ranged from 0.5666 to 0.7041. The balanced accuracy of PART ranged from 0.5041 to 0.6041. The balanced accuracy of the kNN ranged from 0.6708 to 0.7458. The balanced accuracy of SVMLin ranged from 0.6708 to 0.7667. As shown in the figure, the kNN classifier showed the most stable behavior across 30 runs.

Table 8.4 shows the performance measures of the diagnosis of diabetes. Three classifiers obtained a balanced accuracy above 0.70, SVMLin, JRip, and kNN. One of them a rule-based classifier. Accuracies were above 0.70 in all cases. Sensitivity obtained values ranged from 0.8219 to 0.8816, while specificity values ranged from 0.4384 to 0.5746. All kappa values were below 0.50. SVMLin obtained the lowest standard deviation in balanced accuracy.

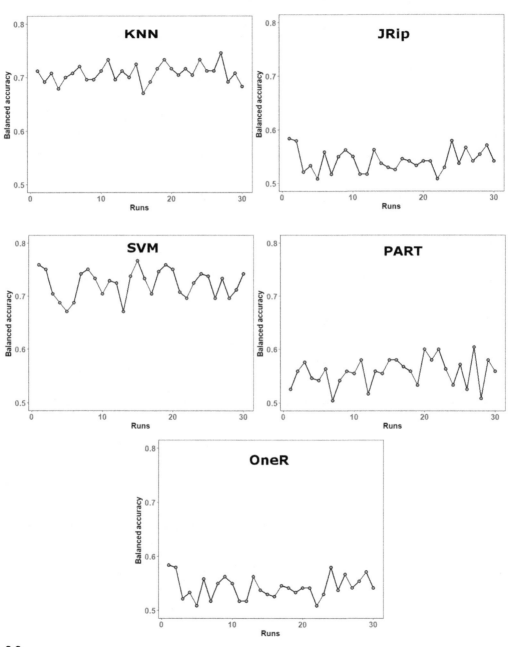

FIG. 8.3

Balanced accuracy across 30 runs for each classifier in hepatitis.

Table 8.4 Average Performance Measures After 30 Runs for Diabetes

Classifier	Balanced Accuracy	Accuracy	Sensitivity	Specificity	Kappa
JRip	0.7100	0.7528	0.8455	**0.5746**	0.4305
(SD)	0.0114	0.0097	0.0172	0.0282	0.0178
kNN	0.7055	0.7584	0.8730	0.5380	0.4337
(SD)	0.0076	0.0062	0.0061	0.0144	0.0153
OneR	0.6513	0.7185	0.8642	0.4384	0.3261
(SD)	0.0074	0.0069	0.0100	0.0141	0.0158
PART	0.6953	0.7353	0.8219	0.5687	0.3990
(SD)	0.0125	0.0112	0.0361	0.0515	0.0215
SVMLin	**0.7225**	**0.7728**	**0.8816**	0.5635	**0.4685**
(SD)	0.0043	0.0035	0.0049	0.0092	0.0085

Note: The highest values appear in bold.
SD, standard deviation.

In Fig. 8.4 we show each of the classifiers with the balanced accuracy as well as the average error rate for diabetes. The SVMLin classifier obtained the best result with a balanced accuracy of 0.7225, while the rest of the classifiers varied between 0.6513 and 0.7100 of balanced accuracy. Again OneR was the classifier with the lowest average result. That is, the SVMLin is the classifier that obtained the most hits with 72.25% in the test cases.

Fig. 8.5 shows the behavior of the balanced accuracy of each of the classifiers across 30 runs for diabetes. The balanced accuracy of OneR ranged from 0.6326 to 0.6630. The balanced accuracy of JRip varied from 0.6883 to 0.7236. The balanced accuracy of PART ranged from 0.6730 to 0.7190. The balanced accuracy of kNN ranged from 0.6914 to 0.7226. The balanced accuracy of SVMLin ranged from 0.7111 to 0.7286. As shown in the figure, the SVMLin classifier showed the most stable behavior across 30 runs.

Table 8.5 shows the performance measures of the diagnosis of Parkinson. This disease obtained the best classification results of all. Four classifiers obtained a balanced accuracy above 0.80; kNN, JRip, SVMLin, and PART. Two of these are rule-based classifiers. Accuracies were above 0.85 in all cases. Sensitivity obtained values ranged from 0.5642 to 0.8900, while specificity ranged from 0.9045 to 0.9736. All kappa values were above 0.60. kNN obtained the lowest standard deviation for the balanced accuracy.

In Fig. 8.6 we show each of the classifiers with the balanced accuracy as well as the average error rate for Parkinson. The kNN classifier obtained the best result with a balanced accuracy of 0.9267, while in the rest of the classifiers the balanced accuracy varied between 0.7689 and 0.8335. OneR was the classifier with the lowest average result. That is, kNN is the classifier that obtained the most hits with 92.67% in the test cases.

Fig. 8.7 shows the behavior of the balanced accuracy of each of the classifiers across 30 runs for Parkinson. The balanced accuracy of OneR ranged from 0.7196 to 0.8053. The balanced accuracy of JRip varied from 0.7642 to 0.8892. The balanced accuracy of PART ranged from 0.7250 to 0.8750. The balanced accuracy of the kNN ranged from 0.8910 to 0.9482. The balanced accuracy of SVMLin ranged from 0.7732 to 0.8553. As shown in the figure, the kNN classifier showed the most stable behavior across 30 runs.

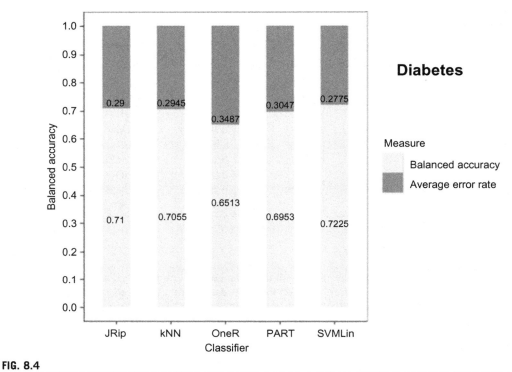

FIG. 8.4

Balanced accuracy and average error rate across 30 runs for diabetes.

Table 8.6 shows the performance measures of the diagnosis of arrhythmia. Two classifiers obtained a balanced accuracy above 0.70, PART and JRip. For the first time, the two best classifiers were rule-based classifiers. Accuracies were below 0.70 in most cases. Sensitivity obtained values ranged from 0.7423 to 0.9264, while specificity from 0.1938 to 0.6555. All kappa values were below 0.50. kNN obtained the lowest standard deviation in the balanced accuracy.

In Fig. 8.8 we show each of the classifiers with the balanced accuracy as well as the average error rate for arrhythmia. PART obtained the best result with a balanced accuracy of 0.7448, that is, PART was the classifier that obtained the most hits with 74.48% in the test cases. In the rest of the classifiers, the balanced accuracy varied between 0.5163 and 0.7138. OneR was the classifier with the lowest average result.

Fig. 8.9 shows the behavior of the balanced accuracy of each of the classifiers across 30 runs for arrhythmia. The balanced accuracy of OneR ranged from 0.4944 to 0.5368. The balanced accuracy of JRip varied from 0.6680 to 0.7486. The balanced accuracy of PART ranged from 0.7229 to 0.7756. The balanced accuracy of the kNN ranged from 0.6020 to 0.6409. The balanced accuracy of SVMLin ranged from 0.6528 to 0.7173. As shown in the figure, kNN showed the most stable behavior across 30 runs.

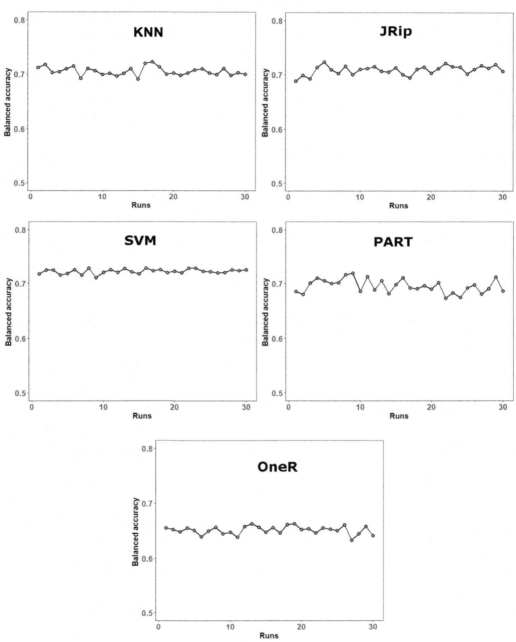

FIG. 8.5

Balanced accuracy across 30 runs for each classifier in diabetes.

Table 8.5 Average Performance Measures After 30 Runs for Parkinson

Classifier	Balanced Accuracy	Accuracy	Sensitivity	Specificity	Kappa
JRip	0.8335	0.8785	0.7525	0.9145	0.6546
(SD)	0.0313	0.0181	0.0648	0.0199	0.0525
kNN	**0.9267**	**0.9470**	**0.8900**	0.9633	**0.8477**
(SD)	0.0126	0.0074	0.0242	0.0064	0.0216
OneR	0.7689	0.8826	0.5642	**0.9736**	0.6125
(SD)	0.0228	0.0131	0.0439	0.0108	0.0453
PART	0.8098	0.8624	0.7150	0.9045	0.6084
(SD)	0.0373	0.0230	0.0709	0.0208	0.0671
SVMLin	0.8115	0.8785	0.6908	0.9321	0.6393
(SD)	0.0187	0.0119	0.0338	0.0097	0.0356

Note: *The highest values appear in bold.*
SD, *standard deviation.*

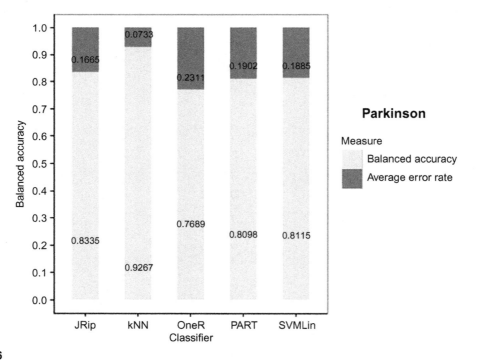

FIG. 8.6

Balanced accuracy and average error rate across 30 runs for Parkinson.

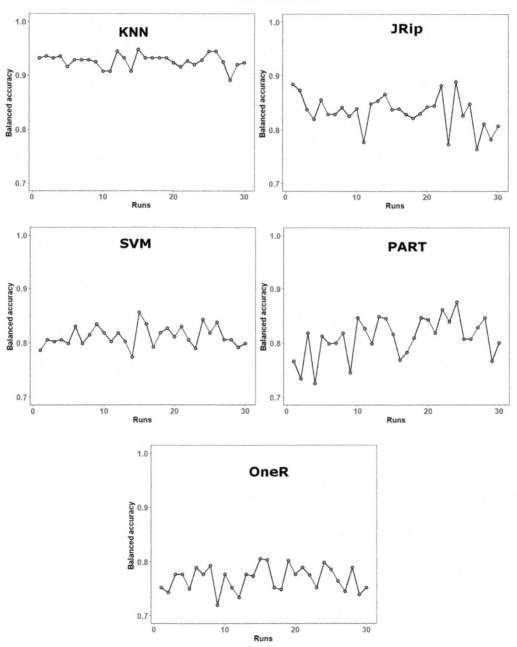

FIG. 8.7

Balanced accuracy across 30 runs for each classifier in Parkinson.

Table 8.6 Average Performance Measures After 30 Runs for Arrhythmia

Classifier	Balanced Accuracy	Accuracy	Sensitivity	Specificity	Kappa
JRip	0.7138	0.7222	0.7722	**0.6555**	0.4299
(SD)	0.0215	0.0207	0.0257	0.0360	0.0428
kNN	0.6268	0.6696	**0.9264**	0.3272	0.2732
(SD)	0.0087	0.0081	0.0088	0.0151	0.0185
OneR	0.5163	0.5623	0.8387	0.1938	0.0352
(SD)	0.0107	0.0110	0.0166	0.0150	0.0233
PART	**0.7448**	**0.7593**	0.8465	0.6431	**0.4990**
(SD)	0.0165	0.0150	0.0224	0.0370	0.0322
SVMLin	0.6863	0.6942	0.7423	0.6301	0.3737
(SD)	0.0136	0.0135	0.0194	0.0216	0.0274

Note: *The highest values appear in bold.*
SD, *standard deviation.*

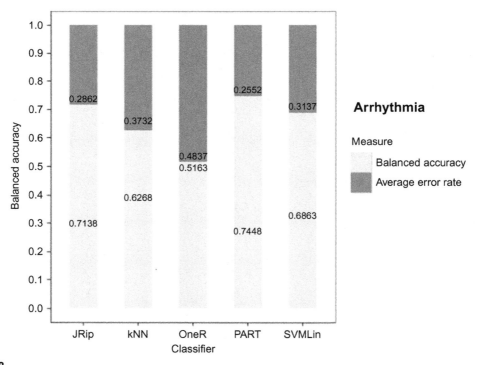

FIG. 8.8

Balanced accuracy and average error rate across 30 runs for arrhythmia.

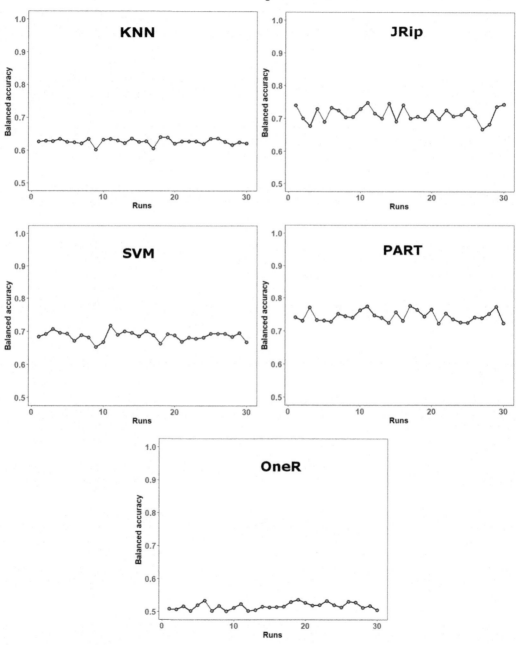

FIG. 8.9

Balanced accuracy across 30 runs for each classifier in arrhythmia.

8.5 CONCLUSIONS

In this study, we compared the behavior of rule-based classifiers against two other classifiers of different approaches, such as kNN and SVMLin, in the diagnosis or prognosis of four diseases: hepatitis, arrhythmia, diabetes, and Parkinson.

Based on the results obtained from the experiments, we conclude that JRip was the best of rule-based classifiers, whereas SVM achieved better results among the most complex classifiers. For diabetes and hepatitis diseases, SVM scored the highest in accuracy, averaging 0.7225 and 0.7232, respectively; for Parkinson the best classifier was kNN with 0.9267 and for arrhythmia it was the PART classifier with averages of 0.7448. SVM and kNN were found to outperform rule-based classifiers by predicting three out of four diseases with the best performance.

It is expected that this study will represent a contribution to disease diagnosis-prognosis area using classifiers. From the medical point of view, it is essential for the specialist to have the appropriate information that allows a correct diagnosis to be made. From computer science, it is important to explore the differences between the different classification algorithms in different domains.

As future work, we propose to:

- use the JRip, OneR, PART, SVM, and kNN classifiers in multiclass classification tasks, for an arrhythmia dataset.
- make prediction models with other classifiers such as the ensemble, since these methods have been shown to achieve better results in classification tasks, according to the published literature.
- use feature selection, since this technique allows an increase in precision results by selecting the best predictors.

REFERENCES

[1] Azar AT, El-Metwally SM. Decision tree classifiers for automated medical diagnosis. Neural Comput Appl 2013;23(7–8):2387–403.

[2] Orru G, Pettersson W, Marquand AF, Sartori G, Mechelli A. Using support vector machine to identify imaging biomarkers of neurological and psychiatric disease: a critical review. Neurosci Biobehav Rev 2012;36(4):1140–52.

[3] Zhang Y, Dong Z, Phillips P, Wang S, Ji G, Yang J, et al. Detection of subjects and brain regions related to Alzheimer's disease using 3D MRI scans based on eigenbrain and machine learning. Front Comput Neurosci 2015;66(9):1–115.

[4] Lichman M. UCI machine learning repository. University of California, Irvine, School of Information and Computer Sciences; 2013. http://archive.ics.uci.edu/ml.

[5] Little MA, McSharry Patrick E, Roberts SJ, Costello DAE, Moroz IM. Exploiting nonlinear recurrence and fractal scaling properties for voice disorder detection. Biomed Eng Online 2007;6(1):23.

[6] Cohen WW. Fast effective rule induction. In: Twelfth international conference on machine learning. San Francisco, CA, USA: Morgan Kaufmann; 1995. p. 115–23.

[7] Rajput A, Aharwal RP, Dubey M, Saxena SP, Raghuvanshi M. J48 and JRip rules for e-governance data. IJCSS 2011;5(2):201.

[8] Witten IH, Frank E, Hall MA, Pal CJ. Data mining: practical machine learning tools and techniques. San Francisco, CA, USA: Morgan Kaufmann; 2016.

[9] Shmueli G, Patel NR, Bruce PC. Data mining for business intelligence: concepts, techniques, and applications in Microsoft Office Excel with XLMiner. Hoboken, NJ, USA: John Wiley and Sons; 2011.

[10] Alpaydin E. Introduction to machine learning. Cambridge: MIT Press; 2014.

[11] Vapnik V. Statistical learning theory. New York: Wiley; 1998.

[12] Mukherjee S. Classifying microarray data using support vector machines. In: A practical approach to microarray data analysisNew York: Springer; 2003. p. 166–85.

[13] Cristianini N, Shawe-Taylor J. An introduction to support vector machines and other kernel-based learning methods. Cambridge: Cambridge University Press; 2000.

[14] Burges CJC. A tutorial on support vector machine for pattern recognition. Data Min Knowl Discov 1998;2:955–74.

[15] Cohen J. A coefficient of agreement for nominal scales. Educ Psychol Meas 1960;20(1):37–46.

[16] Cohen. Statistics how to, statistics for the rest of us!; 2017.

SMART SENSOR APPLICATION FOR HEALTH AND WELL-BEING APPLICATIONS

ASSESSING THE PERCEPTION OF PHYSICAL FATIGUE USING MOBILE SENSING

9

Netzahualcóyotl Hernández[*,†], **Jesús Favela**[†]

Ulster University, Newtownabbey, United Kingdom[*] *CICESE Research Center, Ensenada, Mexico*[†]

9.1 INTRODUCTION

The population worldwide is aging as a result of improvements in health services and lower fertility rates. It is expected that by 2050, 20% of the population worldwide will be older adults [1]. Functional capacity and independence tend to diminish with age due to limited mobility, lose of strength, or other physical or mental problems.

As people age they tend to experience frailty, manifested as health and social challenges, such as decreasing vision, reduced mobility speed, loss of muscle mass, and partial hearing loss. These problems increase the demand for medication, lifestyle counseling, specialized assistance, and care attention.

The increase in chronic and age-related diseases calls for a change from our current emphasis on managing diseases towards an approach aimed at preventing them [2], including tools to infer, monitor, and change behaviors that might hamper well-being.

9.1.1 PHYSICAL FATIGUE

Physical fatigue is an adaptive and regulatory symptom that is perceived by individuals as a need to rest, as a subjective feeling of tiredness [3]. It is experienced as progressive muscle tension, associated with blood irrigation and muscle oxygen supply [4]. The symptom can progress through the following phases: (1) incubation, characterized by the person becoming either nervous, irritated and impatient, or depressed and passive; (2) febricity; with these symptoms worsening; and (3) apathy, characterized by a state of physical or/and psychological problems that might require medical treatment. When these phases are reached repetitively, the muscles and nervous system deteriorate, causing chronic fatigue and symptoms of frailty, a syndrome of increased vulnerability that is associated with a progressive decline in physical and cognitive abilities, and which can eventually lead to an increased risk of fall, fractures, and even death [5].

9.2 ASSESSING THE PERCEPTION OF PHYSICAL FATIGUE

The diagnosis of physical fatigue is conducted using any of the following three methods: a clinical test; completing questionnaires, performing a physical test, and using instruments such as the Borg scale [6]; or by referencing the Workout zone [7]. These methods require the individual to attend a clinic, and thus are not conducted frequently, and they can be unreliable (e.g., when based on self-report). This leads to our interest in developing naturalistic and unobtrusive approaches to detecting the perception of physical fatigue, for instance while the subject performs daily-life activities, such as walking. Even if the method is not as reliable as a test performed in a specialized laboratory, the frequency with which the phenomena is estimated allows for a better assessment of changes in perceived fatigue. If the assessment is performed once every few months there is a risk of obtaining inaccurate measures; i.e., if the test is performed on a day in which the subject is particularly sensitive to the perception of fatigue or otherwise feeling better than normal, the results of the test can be unreliable. In addition, patients and physicians frequently ignore physical fatigue, since it tends to be stereotyped with aging, and thus becomes underdiagnosed and not properly treated [8].

This work presents an approach to unobtrusively monitoring the perception of physical fatigue with the use of everyday technology, namely smartphones. These devices are used to gather physiological data when the individual walks, in order to detect episodes of fatigue [9], as we presented in a preliminary study reported in Ref. [10].

Pervasive technologies have been proposed to identify symptoms of sleepiness by detecting symptoms such as yawning and blinking [11], or changes in the grasping force of a flywheel [12]. Dedicated wearable sensors have also been proposed to estimate physical fatigue, as in the CardioBuddy[1] project that uses optical sensors to measure heart rate [13], or the use of sensors distributed along the body of the patient [14,15]. These approaches require dedicated hardware and might be cumbersome to use or require special assistance, which prevents their daily use.

9.3 AN APPROACH TO ESTIMATE PHYSICAL FATIGUE

The technique we propose avoids the use of dedicated devices. This approach is based on the fact that heart rate rises as the demand for oxygen increases [16], which in turn is perceived by the individual as physical fatigue [7]. Therefore, in order to infer physical fatigue, we estimate the oxygen consumption while the subject walks, by using the accelerometer sensor embedded into mobile devices [17]. We then estimate heart-rate changes calculated based on the oxygen consumption (Fig. 9.1). To estimate motion, we use accelerometer and location (GPS) data collected opportunistically from smartphones, as proposed in Ref. [18].

9.3.1 EXTRACTION

The method begins by obtaining and removing the component associated with gravity from the accelerometer data[2] according to Eq. (9.1).

---

[1]http://www.azumio.com/apps/cardio-buddy-2/.
[2]http://developer.android.com/reference/android/hardware/SensorEvent.html.

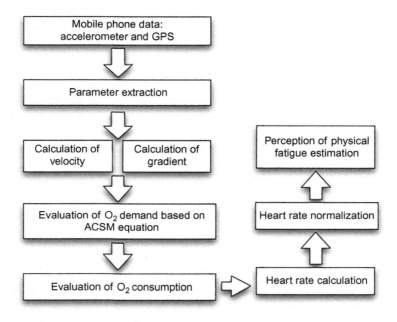

FIG. 9.1

Diagram to describe the approach proposed.

$$A_d = -g - \frac{\sum F}{\text{mass}}$$ (9.1)

where A_d represents the acceleration measured by the sensor, g is the gravity constant of $9.81\,\text{m/s}^2$, and $\sum F$ represents the force applied to the sensor itself.

Afterwards, a sliding window is used to extract the DC element (component direct) under a regular segmentation ($W = 24\,\text{s}$). The data from each window is analyzed to estimate the variables required by the predictive methodology, which consists of walking speed and oxygen demand, as described in the following text.

9.3.2 SPEED

We quantify the number of steps the user walks to measure walking velocity using an estimated stride size, as shown in Eq. (9.2).

$$S_k = D_k/W \quad D_k = ST_k \times SL \quad SL = D_{\text{total}}/ST_{\text{total}}$$ (9.2)

where S represents the velocity (m/s), D represents the distance traveled (m), W is the segment size (s), ST is the total number of steps, and SL is the size of the stride.

9.3.3 GRADIENT

The slope is a physical feature of the terrain, where a gradient value of zero represents horizontality in a landform. It is determined by calculating the user's elevation from the location sensor of the smartphone and the distance walked. Its relevance relies on the impact that transporting, either uphill or downhill, has in a user's physical performance.

9.3.4 OXYGEN DEMAND

The oxygen demand is estimated using the expression proposed by the American College of Sports Medicine (ACSM) [19], and adjusted for each user as shown in Eq. (9.3).

$$
\begin{aligned}
K &= R + H + V \\
R &= 3.6145 - (0.0367 \times \text{BMI}) - (0.0038 \times \text{age}) + (0.1790 \times \text{gender}) \\
H &= 0.1 \times \text{speed} \\
V &= 1.8 \times \text{speed} \times \text{gradient}
\end{aligned}
\tag{9.3}
$$

where R is the user-customized demand for oxygen [17], H is the horizontal element of the displacement speed (m/min), and V is the vertical element relative to the speed (m/min) and gradient displacement (%).

9.3.5 OXYGEN INTAKE

Oxygen demand is estimated periodically as the user walks (as previously described in Eq. 9.2). Oxygen consumption is then estimated, taking into account the variation within intervals of time i along the collected sensor data. For instance, if the oxygen demanded in the interval i_x is greater than the oxygen consumed in the segment i_{x-1}, data will be treated as an increment of effort. In Eq. (9.4) it is shown how this is estimated depending on whether the rate changes, for example, when predicting an increase, a decrease or no change is identified when compared to the previous prediction.

1. When $K_i > V_i$ (increase)

$$\Delta U_i = K_i e^{-\frac{T}{P}} \quad \text{and}$$
$$V_i = V_{i-1} + \Delta U_i$$

2. When $K_i < V_i$ (decrease)

$$\Delta D_i = K_i \left(1 - e^{-\frac{T}{P}}\right) \quad \text{and}$$
$$V_i = V_{i-1} - \Delta D_i$$

3. When $K_i = V_i$ (no change) $\tag{9.4}$
   - When the previous interval's trend increases

$$\Delta U_i = K_i \left(e^{-\frac{T}{2P}} - e^{-\frac{T}{P}}\right) \quad \text{and}$$
$$V_i = V_{i-1} + \Delta U_i$$

   - When the previous interval's trend decreases

$$\Delta D_i = K_i \left(\left(1 - e^{-\frac{T}{2P}}\right) - \left(1 - e^{-\frac{T}{P}}\right)\right) \quad \text{and}$$
$$V_i = V_{i-1} - \Delta D_i$$

where K is obtained from Eq. (9.3); V represents the oxygen consumption of the interval i $(i = 1,2,3\ldots)$; $\Delta U, \Delta D$ represents the increase and decrease of oxygen consumed, respectively; P is the range time to estimate oxygen demand, and T represents the difference between the total time walked by the subject and the time when the oxygen calculation takes place. A predictive model is then generated using a support vector machine (SVM) with the variables just described.

9.3.6 MODEL TO ESTIMATE HEART RATE

As mentioned, this proposal is based on previous work that uses a neutral network (NN) as the classifier [18]. In this project, it has been modified to improve its accuracy and to estimate physical fatigue (as described in Section 9.4). Such modifications include a more accurate estimation of oxygen consumption by customizing the method to the user's profile and the use of SVM as classifier.

WEKA was used to build a predictive model using the SVM algorithm as classifier [20] with a nonlinear kernel. The data used to train the classifier consisted of the smartphone's accelerometer and location data. The accuracy was compared, showing that the lowest validation error (not in the training set) was achieved using polynomial functions, which were then used in the next phases to validate the approach.

9.3.7 NORMALIZATION

Initial results showed that the resulting prediction model produced a result similar to the ground truth, i.e., heart rate (which will be further explained in Section 9.4.2), but with a slight change in position in amplitude. Thus, we normalized the prediction data to correct for the offset.

To normalize we first identify a stabilized-point set to 180 s [21] (Fig. 9.2A), so that the predicted heart rate merges to ground truth (Fig. 9.2B), and then we remove the segment of data corresponding to

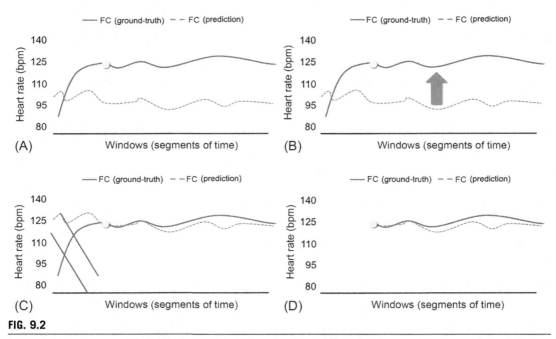

FIG. 9.2

Normalization process: (A) stabilization point, (B) convergence, (C) exclusion, and (D) corrected prediction.

Table 9.1 Relationship of Physical Fatigue Perception Based on Workout Zone Taxonomy [7]

Zone—Description	% HRmax
Very light—similar to walking slowly for a short period of time	50–60
Light—equivalent to working for a long period of time	60–70
Moderate—experienced as difficulty when talking while walking; people are aware of the difficulty in breathing	70–80
High—experienced as deep and forceful breathing when walking at a fast pace. It is not possible to converse in this phase	80–90
Extreme—experienced as a sensation of body heaviness and tiredness	90–100

the aforementioned stabilized point (Fig. 9.2C and D) to keep meaningful data only, as shown in Eq. (9.5).

$$H_i = P_i + (H_{start} - P_1) \qquad (9.5)$$

where H represents the heart rate (postnormalization) window i ($i = 1,2,3...$), P represents the prediction by the mathematical model (prenormalization), and H_{start} is the normalization point, which represents the beginning of the stabilization stage effort of the subject.

9.3.8 PERCEPTION OF PHYSICAL FATIGUE

To estimate the perception of physical fatigue, we first determine the maximum heart rate [22], to then classify the estimated heart rate (postnormalization) using the five Workout zones [7], where the distance between them stands for a margin of 10% of the maximum exercise capacity supported by each participant (Table 9.1).

Notice that, so far, we have addressed the normalization process by assuming that we know the starting heart rate measurement, limiting the model at runtime; we elaborate on this in the discussion section.

9.4 EVALUATION

We evaluated the proposed model with two different populations (i.e., young adults and older adults). We use two smartphones as part of our set-up to gather accelerometer data. The results from both devices (located at the right and left hip) were highly correlated (0.92). Thus, we only report results from the device located on the right hip.

9.4.1 SUBJECTS

We recruited four young adults (two males and two females) with an average age of 28.66 ($\sigma = 2.35$ years), and four older adults (two males and two females) with an average age of 66.5 ($\sigma = 5.8$ years). The participants were preevaluated (completing a questionnaire) to identify whether they could perform aerobic exercises. Exclusion criteria included an impairment to conduct physical exercises, taking medication that could alter the metabolic rate, or a previous diagnosis of cardiovascular abnormalities.

Inclusion criteria consisted of a physical activity history evaluation, with participants successfully completing the Readiness Questionnaire Physical Activity (PAR-Q) [23], a freely mobile rating from Timed Up and Go Test [24], as well as satisfying the 6MWT recommendation (i.e., resting heart rate of <120 bpm) [25]. Physical status was determined by a self-report survey, in which the participants were asked to select a category among five activities that best described their pattern of daily-life activities according to SRPA (Self-Reported Physical Activity index) taxonomy [26].

Participants signed an informed consent form to authorize the use of their gathered data for scientific purposes.

9.4.2 PROCEDURE

Self-reported physical fitness and physiological measurements were obtained on the test day, to guarantee the well-being of each participant. Anthropometric assessment combined weight and height [27] from which the body mass index (BMI) was estimated, to be considered in the ACSM's oxygen demand equation (as discussed in Section 9.3). Subjects were asked to attend the test day following specific criteria for preventing cardiovascular atypical behavior, such as: not having eaten, not having consumed coffee or alcohol 2 h before testing, and not conducting intense exercise for at least 20 min before the test. In addition, we guarantee no heart rate atypical behavior by including 10 min of rest on a bed in a dark, quiet room before beginning any test [28]. Subjects were cooperative during the measurement period, which lasted approximately 15 min.

Before beginning the test, we trained the subjects on the use of a treadmill and equipped them with an electrode beating reader Zephy[3] to track heartbeats; the device gathers data at a speed of 1 Hz (the function of this data was to serve as ground truth), and two smartphones (Samsung Google Nexus S GT-I9020T[4] with three axes accelerometer KR3DM at 50 Hz). Each participant used the electrode beating reader on the chest and the smartphone on the hips. We used a 24-s segment and 50% of overlap.

To collect data, subjects were asked to walk the treadmill following the Naughton physical protocol [29] and a modified protocol version for younger adults (to achieve physical fatigue levels) in a controlled area, i.e., room temperature of 24–25°C as recommended by Scott [28] and a humidity of 62%–64%. We used an electrical pulse reader (Zephy) to monitor the heart rate of the participants as they walked.

A medical doctor was available during the older adults' testing to monitor the stability of the participants during and after the testing performance.

9.4.3 MODEL TRAINING

As discussed in Section 9.3, we trained our model with accelerometer and gradient data (determined according to the Naughton physical protocol), and heart rate measurement (ground truth). We trained

---

[3]http://www.zephyranywhere.com.
[4]http://www.samsung.com.

Table 9.2 Average Postnormalized Tests (T1 and T2) Performed by the Young Adult Population; Subjects YA 1–4

	T1* – T2** (Time)	T2* – T1** (Time)	Error
Sub. 1-YA	6.40 bpm (23.4 min)	2.42 bpm (8.8 min)	4.41 bpm (3.81%)
Sub. 2-YA	8.88 bpm (26 min)	12.90 bpm (9.6 min)	10.89 bpm (9.56%)
Sub. 3-YA	7.44 bpm (25.4 min)	4.58 bpm (12.6 min)	6.01 bpm (5.75%)
Sub. 4-YA	7.62 bpm (25.6 min)	9.30 bpm (13.6 min)	8.46 bpm (8.11%)
Error			7.44 bpm (6.81%)

We trained SVM with T* and predicted with T**.

Table 9.3 Average Postnormalized Tests (T1 and T2) Performed by the Older Adult Population; Subjects AM 1–4

	T1* – T2** (Time)	T2* – T1** (Time)	Error
Sub. 1-OA	13.06 bpm (13 min)	5.36 bpm (9.8 min)	9.21 bpm (8.53%)
Sub. 2-OA	5.33 bpm (8.6 min)	5.66 bpm (8.8 min)	5.44 bpm (4.62%)
Sub. 3-OA	2.23 bpm (19.4 min)	3.30 bpm (17.4 min)	2.77 bpm (2.90%)
Sub. 4-OA	7.51 bpm (13 min)	4.78 bpm (12.6 min)	6.14 bpm (5.73%)
Error			5.89 bpm (5.45%)

We trained SVM with T* and predicted with T**.

our model considering partial data gathered from both participant groups; for instance, we trained our model with data from the first test to predict the second test and vice versa (Tables 9.2 and 9.3).

9.4.4 DATA GATHERING

Each participant completed two tests each, in a period of 2 weeks (June 2014). The average walking duration for young adults was 22 min, and the average time for the elderly population was 17 min. The test was terminated if the subjects achieved 85% of heart rate maximum capacity or if they expressed experiencing physical discomfort, or simply requested to stop the test.

(1) *Younger adults' population*

This population did two tests per participant, the first following the Naughton's protocol and the second using the modified Naughton protocol. The average postnormalized error for these subjects is 7.44 bpm (which represents 6.81% of total error), which coincides with that reported by other works [18]. Table 9.2 shows the results of these tests.

(2) *Older adults' population*

This population did two tests per participant, both following the Naughton protocol. The average postnormalization error for this population is 5.89 bpm (which represent 5.45% of total error), which is better than that reported in the literature [18]. Table 9.3 shows the results of these tests.

This approach shows an improvement of 3.91% and 8.67% in classification error when comparing young adults and older adults, respectively, against those reported previously [18] (Table 9.4).

9.4.5 PERCEPTION OF PHYSICAL FATIGUE INFERENCE

We determined the Workout zone with the percentage of maximum effort from each participant. A high-classification accuracy was obtained with the young population: 90.63%, when comparing physical fatigue based reading (ground truth) against postnormalized prediction (Table 9.5), and 92.04% for the older population (Table 9.6).

Table 9.4 Results From Using Referenced Work Method* and From Using Our Current Approach for Both Populations (i.e., Younger Adults: YA, and Older Adults: OA)**

	NN* (%)	SVM** (%)		NN* (%)	SVM** (%)
Sub. 1-YA	6.23	3.81	Sub. 1-OA	8.49	8.53
Sub. 2-YA	19.68	9.59	Sub. 2-OA	5.96	4.62
Sub. 3-YA	5.20	5.75	Sub. 3-OA	14.24	2.90
Sub. 4-YA	11.76	8.11	Sub. 4-OA	27.77	5.73
Error	10.72	6.81		14.12	5.45

Table 9.5 Physical Fatigue Classification for Younger Tested Subjects 1–4

	T1* – T2**	T2* – T1**	Accuracy (%)
Sub. 1-YA	91.45	100	95.73
Sub. 2-YA	84.62	95.83	90.22
Sub. 3-YA	82.68	100	91.34
Sub. 4-YA	85.16	85.29	85.23
Error	85.98	95.28	90.63

We trained SVM with T to predict T**.*

Table 9.6 Physical Fatigue Classification for Older Tested Subjects 1–4

	T1* – T2**	T2* – T1**	Accuracy (%)
Sub. 1-OA	56.92	91.84	74.38
Sub. 2-OA	100	100	100
Sub. 3-OA	100	100	100
Sub. 4-OA	90.77	96.83	93.80
Error	86.92	97.17	92.04

We trained SVM with T to predict T**.*

The results in the validation phase show that the behavior of the method for estimating the perception of physical fatigue in both groups of participants (i.e., young adults and older adults) is similar and over 90% accuracy for both populations is reported.

9.5 APPLICATION

To illustrate how the methodology proposed can be applied, we conducted an additional study under naturalistic conditions. Subjects used a smartphone application to register when they experienced physical fatigue. Participants were asked to carry the smartphone while walking as part of their daily activities. The four young adults who participated in the previous study were our participants (average age of 28.66, $\sigma = 2.35$ years). Data was collected for over 30 days (from July 7 to August 11, 2014). During this period we recorded 143 walking segments, eliminating 26, since they did not fit the inclusion criteria that was set a priori (i.e., walking for a period of at least 7 min, with full accelerometer and location data).

We analyzed the data gathered when the subjects walked during the day (e.g., distance walked, gradient, perception of physical fatigue, etc.). Results show that our technique could be included as a smartphone application allowing users to monitor their physical condition when recovering from an intervention, or to track their physical deterioration due to aging.

To illustrate how this method could be used, we describe a scenario related to monitoring the physical decline of an older adult, which might signal frailty.

Sara is 67 years, recently had a surgery and was given a new medication to speed her recovery. The physician asked Martha; her daughter, to monitor her progress and the presence of secondary effects such as apathy and fatigue, and recommends them to install the app to monitor fatigue on Sara's smartphone. One morning, after having breakfast, Sara was taking her usual walk in a park nearby, when her daughter received a notification indicating that her mother was having another episode of fatigue, the third one since taking the medication. Martha used the app to send the report to the physician and made an appointment for her mother. After performing clinical tests, the physician decided to change Sara's medication.

As appreciated in previous scenarios, our approach could be used independently or combined with other parameters (e.g., food, quality of sleep) to provide a more comprehensive overview of the user's health in order to be used as an assist tool to constantly monitor health status.

9.6 DISCUSSION

Results of the study show that by personalizing the oxygen consumption equation [19] according to each participant's profile and using SVM we were able to obtain a reliable prediction model, as described in this paper. We also show that one walking test and a 3-min period (as stabilization sample) to train the model could be sufficient to train our proposed model.

While our results provide evidence of adequate detection of fatigue when individuals perform daily-life activities (under naturalistic conditions) such as walking, environmental conditions such as weather, mood, food/ingested, or medicine can be manifested as atypical cardiovascular behavior [28]. It has been estimated that mental stress can change physical performance up to 15% [30]. An additional challenge in the design of a monitoring mobile app (as the one described here) is the requirement of determining a starting heart rate point, which can then be used to normalize the prediction as described. Thus, we face the challenge of integrating these variables into a more versatile prediction model.

9.7 CONCLUSIONS

We have reported a method aimed at estimating the perception of physical fatigue by predicting heart rate through smartphones; the technique is nonintrusive and can be used under naturalistic conditions. The proposed technique is based on a previous work that reports an average error of 7 bpm [18]. To determine the perception of physical fatigue, the maximum physical capacity and Workout taxonomy are used; where each zone appoints to a margin of 10% of exercise intensity carried by each participant in this study.

The proposed method is based on a prediction model obtained by training a Support Vector Machine. It has been tested with data collected from two user-controlled studies with different populations: four young adults, and four older adults. Results show a similar behavior between both groups of participants with a maximum error of 7.44 and 5.89 bpm, which represent an error of 6.81% and 5.45%; respectively, and a precision of the perception of physical fatigue classification of over 90% compared to monitored heart rate (ground-truth) for both populations.

The perception of physical fatigue frequently foreshadows conditions like cancer, multiple sclerosis, arthritis, renal disease, and HIV infection, among other diseases [31–33]. The approach we have presented could be an auxiliary tool for initial diagnosis and the early detection of diseases, to monitor posttreatment evolution, and to help assess the risk of certain medications on patients.

REFERENCES

[1] World Health Organization. World report on ageing and health report, 2015. Available from:http://apps.who.int/iris/bitstream/10665/186463/1/9789240694811_eng.pdf.

[2] Dishman E. In: Inventing wellness systems for aging in place. Computer (Long Beach Calif) 2004;37 (5):34–41.

[3] Singleton WT. Deterioration of performance on a short-term perceptual-motor task. Symposium on fatigue, vol. 8. Oxford: H. K. Lewis & Co.; 1953. p. 163–72.

[4] Rodriguez DR, Fajardo JT. Prevencion de lesiones en el deporte/prevention of sports injuries: claves para un rendimiento deportivo optimo/keys to optimal athletic performance [Spanish edition]. Editorial Medica Panamericana; 2014.

[5] Alvarado BE, Zunzunegui MV, Beland F, Bamvita JM. Life course social and health conditions linked to frailty in Latin American older men and women. J Gerontol A Biol Sci Med Sci 2008;63(12):1399–406.

[6] Borg G. Psychophysical scaling with applications in physical work and the perception of exertion. Scand J Work Environ Health 1990;16:55–8.

[7] Edwards S, Snell M, Sampson E. Sally Edwards' heart zone training: exercise smart, stay fit, and live longer. Massachusetts, USA: Adams Media Corporation; 1996.

[8] Gambert SR. "Why do I always feel tired?" Evaluating older patients reporting fatigue. Consultant 2013;53 (11):785–9.

[9] Vathsangam H, Schroeder ET, Sukhatme GS. Hierarchical approaches to estimate energy expenditure using phone-based accelerometers. IEEE J Biomed Health Inform 2014;18(4):1242–52.

[10] Hernández N, Favela J. Estimating the perception of physical fatigue among older adults using mobile phones. In: Salah AA, Kröse BJA, Cook DJ, editors. Proceedings of the 6th International Workshop on Human Behavior Understanding—Volume 9277. New York, NY, USA: Springer-Verlag, New York, Inc., 2015. p. 84–96

[11] Mohd Noor H, Ibrahim R. Fatigue detector using eyelid blinking and mouth yawning. In: Bolc L, Tadeusiewicz R, Chmielewski L, Wojciechowski K, editors. Computer vision and graphics. Lecture notes in computer science, vol. 6375. Berlin, Heidelberg: Springer; 2010. p. 134–41. https://doi.org/10.1007/978-3-642-15907-7_17.

[12] Baronti F, Lenzi F, Roncella R, Saletti R. In: Distributed sensor for steering wheel rip force measurement in driver fatigue detection. Design, automation test in Europe conference exhibition, 2009 DATE '09; 2009. p. 894–7.

[13] Scully C, Lee J, Meyer J, Gorbach AM, Granquist-Fraser D, Mendelson Y, et al. Physiological parameter monitoring from optical recordings with a mobile phone. IEEE Trans Biomed Eng 2012;59(2):303–6.

[14] Mokaya F, Nguyen B, Kuo C, Jacobson Q, Zhang P. In: [MARS] a real time motion capture and muscle fatigue monitoring tool. Proceedings of the 10th ACM conference on embedded network sensor systems (SenSys '12). New York, NY: ACM; 2012. p. 385–6. https://doi.org/10.1145/2426656.2426721.

[15] Altini M, Penders J, Vullers R, Amft O. Estimating energy expenditure using body-worn accelerometers: a comparison of methods, sensors number and positioning. IEEE J Biomed Health Inform 2015;19(1):219–26.

[16] Corbin CB, Welk GJ, Corbin WR, Welk KA. Concepts of fitness and wellness: a comprehensive lifestyle approach. 9th ed. New York: McGraw-Hill; 2011.

[17] Barstow TJ, Molé PA. Linear and nonlinear characteristics of oxygen uptake kinetics during heavy exercise. J Appl Physiol 1991;71(6):2099–106. Available from, http://www.ncbi.nlm.nih.gov/pubmed/1778898.

[18] Sumida M, Mizumoto T, Yasumoto K. In: Estimating heart rate variation during walking with smartphone. Proceedings of the 2013 ACM international joint conference on pervasive and ubiquitous computing (UbiComp '13). New York, NY: ACM; 2013. p. 245–54.

[19] American College of Sports Medicine. ACSM's metabolic calculations handbook. Baltimore, MD: LWW; 2006.

[20] Hall M, Frank E, Holmes G, Pfahringer B, Reutemann P, Witten I. The WEKA data mining software: an update. SIGKDD Explor 2009;11(1):10–8.

[21] Jeyaranjan R, Goode R, Duffin J. Changes in respiration in the transition from heavy exercise to rest. Eur J Appl Physiol Occup Physiol 1988;57(5):606–10.

[22] Tanaka H, Monahan KD, Seals DR. Age-predicted maximal heart rate revisited. J Am Coll Cardiol 2001;37(1):153–6.

[23] Thomas S, Reading J, Shephard RJ. Revision of the physical activity readiness questionnaire (PAR-Q). Can J Sport Sci 1992;17(4):338–45.

[24] Podsiadlo D, Richardson S. The timed "Up & Go": a test of basic functional mobility for frail elderly persons. J Am Geriatr Soc 1991;39(2):142–8. Available from, http://www.ncbi.nlm.nih.gov/pubmed/1991946.

[25] Enright PL. The six-minute walk test. Respir Care 2003;48(8):783–5.

[26] Jurca R, Jackson AS, LaMonte MJ, Morrow JR, Blair SN, Wareham NJ, et al. Assessing cardiorespiratory fitness without performing exercise testing. Am J Prev Med 2005;29(3):185–93.

[27] Lohman TG, Roche AF. Anthropometric standardization reference manual. Champaign, IL: Human Kinetics Books; 1988.

[28] Scott CB. Resting metabolic rate variability as influenced by mouthpiece and noseclip practice procedures. J Burn Care Rehabil 1993;14(5):573–7. Available from, http://www.scopus.com/inward/record.url?eid=2-s2.0-0027670831&partnerID=40&md5=6b4dc9e6a27ea9ad74c2e5ca47072cf0.

[29] Devis J, Peiro C. La actividad fisica y la promocion de la salud en ninos/as y jovenes: la escuela y la educacion fisica. Revista Psicol del Deport 1993;1:71–86.

[30] Marcora SM, Staiano W, Manning V. Mental fatigue impairs physical performance in humans. J Appl Physiol 2009;106(3):857–64.   [cited 2013 Aug 10]. Available from, http://www.ncbi.nlm.nih.gov/pubmed/ 19131473.

[31] Davis MP, Khoshknabi D, Yue GH. Management of fatigue in cancer patients. Curr Pain Headache Rep 2006;10:260–9.

[32] Braley TJ, Chervin RD. Fatigue in multiple sclerosis: mechanisms, evaluation, and treatment, Sleep 2010;33 (8):1061–7.   Available from, http://www.pubmedcentral.nih.gov/articlerender.fcgi?artid=2910465& tool=pmcentrez&rendertype=abstract.

[33] Adinolfi A. Assessment and treatment of HIV-related fatigue. J Assoc Nurses AIDS Care 2001; 12(Suppl):29–34 [quiz 35-8].

APPLICATIONS TO IMPROVE THE ASSISTANCE OF FIRST AIDERS IN OUTDOOR SCENARIOS

10

Enrique Gonzalez, Raul Peña, Alfonso Avila, David Munoz-Rodriguez
Monterrey Institute of Technology and Higher Education, Monterrey, Mexico

10.1 INTRODUCTION

Information and telecommunications technologies as an area of specialization in engineering currently play a vital and paramount role in human beings' lives. Technology evolution is a steady fact and human beings are testifying to it every day. Day by day, this process is changing our globe and how we interact with each other. Health services are attempting to harness this evolution to benefit the population; mobile health or "mHealth" is a component of eHealth referring to the delivery of healthcare services remotely through mobile devices [1]. MHealth refers to the combined use of information and communications technology; e.g., physiological signals and medical video images are information commonly found in mHealth services, wirelessly transmitted between two different locations. Physiological signals represent the electrochemical changes from the human body, and the most representative are: blood pressure, electrocardiograms (ECGs), body temperature, pulse rate, and respiration rate. These vital signs always provide general information about the status for clinical diagnosis in patients [2].

m-Health promises to be a present and future solution in pursuit of better medical practices in different medical emergency scenarios, such as traffic accidents and illnesses derived from chronic degenerative diseases. People are living longer and the population of our planet is becoming older. Improvements in medical diagnosis and treatments have contributed to the longer life expectancy, as illustrated in Table 10.1. Life expectancy continues to increase in different regions of the world, resulting in a higher demand for better health services. The older population often develops chronic diseases such as diabetes, high-blood pressure, and cardiovascular disorders, among others. People suffering from at least one of these diseases need to continue a normal life and engage in their normal daily outdoor activities. Other emergency scenarios involve unexpected events such as hurricanes, explosions, traffic accidents, etc. These scenarios under outdoor conditions often take place far from medical facilities. These situations demand swift, accurate, and prompt assistance to diminish the damage suffered by injured patients. An adequate and prompt response after an accident can minimize the possibilities of permanent injury or even death.

Intelligent Data Sensing and Processing for Health and Well-being Applications. https://doi.org/10.1016/B978-0-12-812130-6.00010-X

Table 10.1 Life Expectancy Around the World [3]

Country	Life Expectancy		
	Years		
	1990	2009	2015
United States	75	79	79
Spain	77	82	83
Mexico	71	76	77
Japan	79	83	84
Brazil	67	73	75
Tunisia	70	75	75
Australia	77	82	83

The development of new schemes and applications making use of different technologies surrounding mHealth is a key alternative to improving the health services needed to provide adequate and prompt response [4]. The use of wireless body sensor networks (WBSNs) is a key technology for providing this fast response. The deployment of WBSNs enables interaction among caregiver personnel, smart vital-sign sensors, smart devices such as tablets, cell phones, or PDA wireless technology for remote medical assistance. Because of the combination of these elements, it is possible to collect and transmit medical data from patients to a medical center, making real-time monitoring from the accident location possible. WBSN-based applications acquire medical data, such as images, physiological data, and video, from biomedical sensors and cameras. Physicians at the medical center are then able to use this medical data for better and faster diagnosis as well as early treatment during the prehospital assistance period.

10.1.1 TECHNOLOGY FOR MEDICAL EMERGENCIES

Emergency situations can occur due to unexpected events that endanger humans' life. Examples of these unexpected and frequent events are: heart attacks, explosions, natural disasters, traffic accidents, and acts of vandalism [2]. Table 10.2 shows the leading causes of deaths in the United States. These leading causes of death have the potential to trigger emergencies. Assisting at an emergency within the following 60 minutes is key to reducing sequels and risk of death. Adams Cowley, a military surgeon, introduced this "golden hour" concept and he indicated: "any citizen suffering serious traumatic injuries has a 60 minute time window to survive." Additionally, in Ref. [5] the authors pointed out that 75% of traffic deaths are concentrated during the golden hour. Technology can surely make a difference within this golden hour with a platform capable of directly reporting the patient's condition. This platform lets the physician track the situation and be prepared to act upon patient arrival.

As we can see in Table 10.2, besides accidents, diseases like heart attacks, chronic lower respiratory failure, Alzheimer's, and diabetes mellitus are the main causes of deaths in the United States; these chronic degenerative diseases could trigger an emergency event and mHealth systems based on WBSNs could be used to assist.

Table 10.2 The Top 10 Leading Causes of Death in 2013 [5a]	
1	Diseases of heart (heart disease)
2	Malignant neoplasms (cancer)
3	Chronic lower respiratory diseases
4	Accidents (unintentional injuries)
5	Cerebrovascular diseases (stroke)
6	Alzheimer's disease
7	Diabetes mellitus (diabetes)
8	Influenza and pneumonia
9	Nephritis, nephrotic syndrome, and nephrosis (kidney disease)
10	Intentional self-harm (suicide)

10.2 VITAL SIGNS AND PATIENT MONITORING

The elements of and context for physiological monitoring and mHealth infrastructure for emergency scenarios are addressed in this section. It is relevant to have an overview of what kind of information is required from patients and the deployment of a remote system that tracks emergency situations. We discuss these issues briefly.

10.2.1 VITAL SIGNS

Vital signs are measures reflecting the performance of body functions. These numbers are called "vital" because they bring critical information about patients' health conditions. With these measurements, it is possible to identify a medical risk and its magnitude based on and compared with the "normal range." Vital signs can also be markers of chronic degenerative disease status [6].

These parameters are often obtained by first aiders, but also by the patients themselves. Caregivers act depending on metrics obtained during the process: that is, if a risk of emergency is detected, patients could be referred to a medical center. For example, before a heart attack, some parameters such as blood pressure or pulse rate may change, and if this situation is detected in time, the heart attack may be avoided. The most representative vital sign parameters are shown in Table 10.3.

Table 10.3 Most Common Vital Sign Parameters Collected During an Emergency Situation [7]	
Vital Sign	**Description**
Electrocardiography (ECG)	An electrical recording of the heart signal to detect anomalies
Electroencephalography (EEG)	The electrical recording of the brain's activity
Pulse	Measurement of the heart rate, that is, the number of times the heart beats during a determined period
Respiratory rate	Inspiration and expiration activity within a specific time interval
Blood pressure	Pressure of the heart pushing blood, to distribute it against artery resistance
Temperature	Measure of the body's ability to generate and get rid of heat
Oxygen saturation	Measure of the oxygen levels in the blood

10.2.2 EXAMPLES OF DISEASES IN WHICH MHEALTH APPLICATIONS MAY BENEFIT PATIENTS

Health monitoring is very important in terms of prevention, particularly if the early detection of diseases can reduce suffering and medical costs. The diagnosis and prompt treatment of various diseases can radically improve alternatives for the medical treatment of the patient. This is particularly true in cardiovascular diseases and diabetes; the use of sensors can identify patients at risk by monitoring and transmission of vital signs to medical professionals, who may subsequently determine the actions needed to safeguard the patient's health. Table 10.4 shows examples of diseases in which treatment and adequate monitoring could benefit the doctor and the patient for improved health management.

Table 10.4 Parameters Commonly Used to Monitor Different Diseases Where Patients Could Benefit From Continuous Monitoring [8]

Disease Process	Physiological Parameter (BSN Sensor Type)	Biochemical Parameter (BSN Sensor Type)
Hypertension	Blood pressure (implantable/wearable mechanoreceptor)	Adrenocorticosteroids (implantable biosensor)
Ischemic heart disease	Electrocardiogram (EKG), cardiac output (implantable/wearable EKG sensor)	Troponin, creatine kinase (implantable biosensor)
Cardiac arrhythmias/ heart failure	Heart rate, blood pressure, EKG, cardiac output (implantable/wearable mechanoreceptor)	Troponin, creatine kinase (implantable biosensor)
Cancer (breast, prostate, lung, colon)	Weight loss (body fat sensor) (implantable/ wearable mechanoreceptor)	Tumor markers, blood detection (urine, feces, sputum), nutritional albumin (implantable biosensor)
Asthma/COPD	Respiration, peak expiratory flow, oxygen saturation (implantable/wearable mechanoreceptor)	Oxygen partial pressure (implantable/wearable optical sensor, implantable biosensor)
Parkinson's disease	Gait, tremor, muscle tone, activity (wearable EEG, accelerometer, gyroscope)	Brain dopamine level (implantable biosensor)
Alzheimer's disease	Activity, memory, orientation, cognition (wearable accelerometer, gyroscope)	Amyloid deposits (brain) (implantable biosensor/EEG)
Stroke	Gait, muscle tone, activity, impaired speech, memory (wearable EEG, accelerometer, gyroscope)	
Diabetes	Visual impairment, sensory disturbance (wearable accelerometer, gyroscope)	Blood glucose, glycated hemoglobin (HbAlc) (implantable biosensor)
Rheumatoid arthritis	Joint stiffness, reduced function, temperature (wearable accelerometer, gyroscope, thermistor)	Rheumatoid factor, inflammatory and autoimmune markers (implantable biosensor)
Renal failure	Urine output (implantable bladder pressure/ volume sensor)	Urea, creatinine, potassium (implantable biosensor)
Vascular disease (peripheral vascular and aneurisms)	Peripheral perfusion, blood pressure, aneurism sac pressure (wearable/implantable sensor)	Hemoglobin level (implantable biosensor)
Infectious disease	Body temperature (wearable thermistor)	Inflammatory markers, white cell count, pathogen metabolites (implantable biosensor)

10.3 **PREHOSPITAL EMERGENCY ASSISTANCE**
10.3.1 **PREHOSPITAL ASSISTANCE PERSONNEL**

Personnel participating in a prehospital assistance scenario must have adequate training for the attendance of emergency situations. Classifications and protocols are previously established regarding prehospital caregiver capabilities and training. Some of the most representative are first responder, basic, and advanced prehospital trauma care [9]. Each of the classifications play different roles and their collaboration forms the backbone of the emergency medical system to safeguard injured patients' lives.

10.3.1.1 First responders
First responders are people with basic knowledge of what to do in emergencies. First responders are usually present or nearby when an accident occurs, and they must follow these tasks: assess the scene, call for help, assess the victim, provide choking support, provide breathing support, immobilize the victim, provide cardiopulmonary resuscitation if needed, drag the patient to a safe place, and stop any bleeding. At this level individuals are equipped only with basic supplies such as bandages, gloves, immobilizers, etc.

10.3.1.2 Basic prehospital trauma care
These individuals are formally prepared and have a consistent background in patient transportation, scene management, rescue, and stabilization. Based on their knowledge and training, they should be able to evaluate injured or ill patients to decide whether they really need to be transported to a specialized medical center or can be treated at the scene. The scope of practice of these providers includes extrication and rescue, immobilization, and the administration of oxygen. Ambulance services customarily use personnel trained at this level.

10.3.1.3 Advanced prehospital trauma care
These people are commonly physicians or paramedics sufficiently prepared through specialized courses of thousands of hours for emergency assistance. They must have theoretical and practical training which lets them face serious emergency cases in crisis situations. This level of prehospital assistance is not often found worldwide; first-world countries commonly have this level. Within the capabilities available at this level are: endotracheal intubation, needle decompression of a pneumothorax, cricothyroidotomy, and sutures in tissues [9].

10.3.2 **CURRENT PRACTICES FOR PREHOSPITAL EMERGENCY ASSISTANCE**

Prehospital emergency care starts when bystanders or patients themselves request medical assistance by calling an emergency number asking for help. Once first aiders arrive on the scene, they start patient evaluation to obtain an overview of the patient's medical condition. Then, first aiders prepare patients for transfer to a hospital or clinic. During the transfer, first aiders continuously monitor the patient and conduct the necessary medical procedures to safeguard the victim's life. At the hospital, first aiders must provide complete verbal and written information of the time, place, and mechanism of the accident, and also their evaluation of the patient, including signs and symptoms, procedures performed, and evolution during the transfer. Commonly, the job of the prehospital care provider ends after the delivery of a live patient to a hospital, when care is handed over to hospital personnel. Prehospital caregiver guidelines are given in Fig. 10.1 [10].

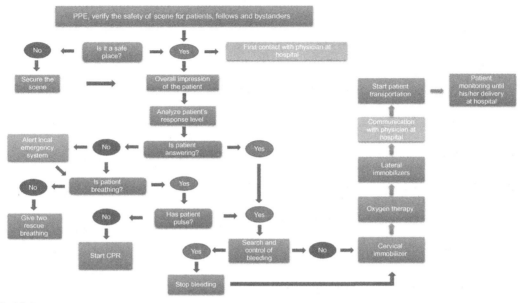

FIG. 10.1

Prehospital assistance panorama for emergency scenarios.

In Fig. 10.1 we can observe that communication between caregivers on the emergency scene and physicians is poor (takes place only twice), and patient medical status is not reported in real time. The communication is done via two way-voice radio systems and the cellular telephone network. The latter is the predominant technology for emergency medical services (EMS) communications with the hospital; however, radio communication systems are the most reliable technology when emergency scenes are located in remote communities [11]. Clear and accurate communication is a crucial factor within medical practice; lack of communication could lead to medical diagnostic errors and subsequently compromise a patient's safety [12]. In pursuit of continuous, accurate, and real-time communication between EMS and the hospital, during the next section we will introduce the concept and layout of an integral mHealth system.

10.4 WBSNS ARCHITECTURE BASED ON M-HEALTH SYSTEMS
10.4.1 WIRELESS BODY SENSOR NETWORK

A WBSN is a completed concept created to provide real-time and continuous monitoring of the human body and its contextual states. WBSNs serve as the integration of mHealth with leading technology and healthcare personnel aiming to bring a better quality of service for people who need primary care. Miniaturized, invasive or noninvasive sensors are responsible for vital signs acquisition, while smartphones and wireless communication technologies allow collection and transmission of the parameters collected from patients [13].

10.4.2 **WBSN ARCHITECTURE**

WBSNs typically consist of multiple sensor nodes worn on the body, each capable of sampling, processing, and communicating wirelessly. The processed healthcare information could be one or more physiological measurements or environmental parameters. The physiological parameters are typically blood glucose and oxygen levels, pulse rate, blood pressure, etc. Antenna components embedded in the sensor nodes make it possible for the data generated to be transmitted wirelessly to a body-worn or closely located hub device, eliminating the need for cables. The hub device, in turn, receives the data generated from the various sensor nodes on the body and may process the data locally and transmit it wirelessly via an appropriate radio link for centralized processing, display and storage, or both [14].

In terms of system architecture, WBSNs are divided into three imaginary layers, as shown in Fig. 10.2: (A) intra-WBSN communications, (B) inter-WBSN communications, and (C) beyond-WBSN communications.

10.4.2.1 Intra-WBSN

This layer considers communication between sensors and personal servers (usually less than 10 m away). The sensor nodes have direct wireless communication (Bluetooth, ZigBee, etc.) with the personal server (mobile device). The mobile device has the capacity to take a video of the patient's condition, collect and/or process the physiological measurements, and determine the patient's location. Also, the mobile device can directly transmit via a mobile network (3G, 4G LTE) all the collected information to the

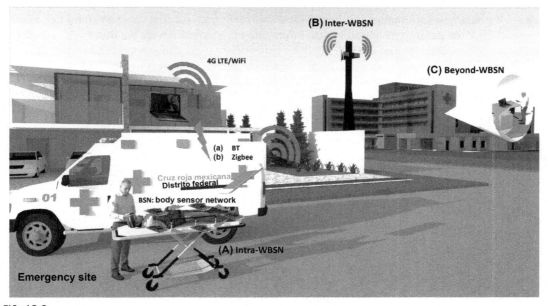

FIG. 10.2

Wireless monitoring platform for emergency situations outdoors: (A) intra-WBSN communications; (B) inter-WBSN communications; (C) beyond-WBSN communications [2].

second layer; but if the device does not have this technology, it can transmit the information to an ambulance via Bluetooth, and later the ambulance can rebroadcast the information [2].

10.4.2.2 Inter-WBSN

This layer considers communication between the personal servers to the Internet. The mobile device can directly transmit the information (3G, 4G LTE, Wi-Fi, WiMAX) to the Internet or cloud services, or indirectly, via one or more relay points, and then to the Internet. In some cases, this is the final layer of the WBSN system, in which the personal servers become the base station to locally process or display the health condition of the patient.

10.4.2.3 Beyond-WBSN

This layer considers communication between the Internet to the local data server. The collected information can be stored in a local data server (medical center), and/or can be sent directly to a personal device (mobile device of the specialist). This is the final layer, which facilitates the real-time medical attention via video conference or mobile calls during an emergency. With the help of a database server, all patient data collected can be stored, displayed, and fully processed to create the patient health profile, which can then be consulted at any time needed.

10.4.3 WBSN INFRASTRUCTURE CONNECTIVITY

Infrastructure connectivity refers to the communication technology implemented between each tier of the mHealth system. Traditional technology included the available communication infrastructures to manage the data flow information in a hierarchical approach, from the first tier to the third tier. Wired or wireless sensors transmit the physiological information to a mobile device via Bluetooth or ZigBee; the mobile device sends the data through a cellular network or a local area network; and finally, the data is managed, visualized, and stored in a local or a remote server, as shown in Fig. 10.3.

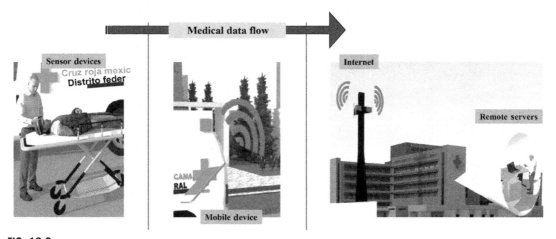

FIG. 10.3

Hierarchical approach for traditional connective technology in a WBSN architecture.

10.5 M-HEALTH SYSTEMS FOR OUTDOORS: RECENT APPLICATIONS AND IMPLEMENTED ARCHITECTURES

10.5.1 M-HEALTH SYSTEMS

These systems have begun to arise as a powerful technology in the search for improved medical assistance for emergency situations. MHealth systems follow the architecture and infrastructure of WBSNs and, over the last decade, they have been widely developed. The feasibility of these systems is due to nearly everyone having access to a mobile phone; the tremendous smartphone evolution with integrated computers and the ability to run software applications is the reason for its acceptance as a major tool to provide medical assistance. Smartphones are also embedded with GPS, Wi-Fi, and Bluetooth (BT) sensors. These features allow swift patient location and connectivity with sensors attached to human bodies. Sensors can have direct wireless communication with smartphones through BT. Additionally, smartphones have the capacity to take video of the patient's condition, compress and encode signals into video, and transmit via Wi-Fi (modem enabled inside an ambulance) or 4G LTE technologies. Moreover, smartphones are not the only technology necessary for mHealth services; the multidisciplinary areas of electrical engineering, computer science, biomedical engineering, and medicine all complement the use of smartphone technology to the benefit of today's healthcare [15].

Through mHealth platforms, doctors can manage adequate actions and treatments to benefit their patients. Nevertheless, many issues concern not just human patients, but network deployment also plays a crucial role in offering a reliable, efficient, and context-aware healthcare system. Patient monitoring in real time must be dynamic and adaptive because patients can be monitored in both immobile or mobile situations. Commonly, immobile monitoring is related to indoor scenarios, as in a clinic or hospital, while mobile situations are within the context of chronic diseases and sport tracking, that is, in outdoor scenarios. The monitoring, depending on the patient's condition, could be routine or emergency [16]; for example, a patient after a surgery will require periodic monitoring to be sure of ongoing medical status, but if the patient presents sudden changes in physiological parameters, an emergency monitoring set-up must be implemented; time intervals of monitoring will be shorter, and some vital signs could acquire more priority than others.

10.5.2 APPLICATIONS REGARDING M-HEALTH SYSTEMS IN THE LAST YEARS

10.5.2.1 Emergency applications for outdoors

Consider a WBSN with sensors reporting data remotely to a medical center through a mobile device, such as a smartphone with video and GPS capabilities. Under a stable health condition, the personal health information (PHI) of the patient should be reported to the medical center once in a while, keeping track of the patient's health status. However, occurrence of an emergency event requires the reporting of PHI, video, location, and alarms in real time. The scenario just introduced is an emergency assistance scenario, capable of immediately responding after the occurrence of the incident, with different technological challenges, including mobile computing, medical sensors, video technologies, location algorithms, and communication for mobile healthcare systems.

Table 10.5 shows emergency scenarios for mHealth applications deployed outdoors. As the applications directly transmit the physiological data or the generated alarms to the medical center to inform them of the location and situation of the patient, to request for emergency services, the applications

Table 10.5 Emergency Context of mHealth Applications Attending Emergencies in Outdoors

Refs.	Emergency Context	Target Location	First Aider Friendly
[17]	If the mCarer system detects a critical situation, it sends a message to the location server and to the emergency service to inform them of the location of the patient	Outdoors	✓
[18]	An alarm is sent if any hazardous condition appears or any of the defined user thresholds are surpassed. The monitoring system is deployed in a sports environment	Outdoors	
[19]	After detection of a fall or cardiac anomaly, the alerts enabler system automatically contacts the patient's family or emergency services	Outdoors	✓
[20]	The system continuously monitors the patient's health condition and if an emergency is detected, a request for ambulance is sent	Outdoors	✓

could be used by first aiders. The first aiders would connect and manage the sensor devices, ensuring that the location and data is correctly transmitted, and wait for the emergency responders.

10.5.2.2 Emergency applications for indoors and outdoors

The applications in Table 10.6 are intended to be deployed for both locations: indoors and outdoors. While at an indoor location, the mHealth system configuration could be stationary, e.g., the sensor nodes are connected wirelessly to a personal computer (PC) connected every time to Internet in the home of the patient; in outdoors locations the mHealth platform must be mobile and capable of automatically locate the person in remote areas. Kim et al. [31] consider an indoor/outdoor scenario to deploy their system. The indoor patient with an advanced condition such as a stroke, paralysis, cardiovascular disease, or diabetes who primarily stays in the hospital is monitored to keep track of and stabilize his condition; after the patient is stable and is sent home for postcare, the WBSN stills tracks his physiological parameters on the way home in case of an emergency occurrence.

As the platforms have different purposes, the target user also varies. The applications in Refs. [24,30] are not suitable for first aiders, because the system is intended to be used by the person. The purpose of the project is to detect falls automatically to provide support to the injured people, so the system must be carried by the person. Another example is the multiple camera system in Ref. [23], in which the transmission of multiple videos and specialized medical devices inside the ambulance need to be managed by the emergency responders, such as paramedics.

10.5.3 IMPLEMENTED ARCHITECTURES IN APPLICATIONS FOR MHEALTH SYSTEMS

10.5.3.1 Emergency architectures for outdoors

As the emergency context is different for each application, the WBSN architecture also has different configurations. Actually, WBSN architecture developments are based on the purpose of the project [32]. Table 10.7 shows the WBSN architectures for each application in Table 10.5. As an example, the second application only stores and visualizes the health information in a local database; the remote server is not addressed [18]. For this scenario, if there is an alert in some physiological parameter of the patient, the alert is only attended to by the emergency center and the patient needs to wait for ambulance

Table 10.6 Emergency Context of mHealth Applications Attending Emergencies for Indoors and Outdoors

Refs.	Emergency Context	Target Location	First Aider Friendly
[21]	If an abnormal condition is detected such as critical cardiac event, patients are referred to the emergency department	Indoors/outdoors	✓
[22]	In a web portal to real-time monitoring of health status of the elderly, their families or clinicians can visualize the person's location and status of alarm of abnormal physiological parameters	Indoors/outdoors	
[23]	Multiple real-time video streams (external and internal) of injured persons that need immediate medical assistance are collected/processed by the ambulance. The videos and vital signs are transmitted to the remote hospital	Indoors/outdoors	
[24]	Alerts are sent in the case of fall detection via SMS and GPS location. The system uses a smartphone to provide the elderly with a unobtrusive monitoring	Indoors/outdoors	
[25]	A context-aware smart solution for people with disabilities to track and determine user's location at their workplace. If there is a critical health issue, a message will be sent to a caregiver	Indoors/outdoors	✓
[26]	If emergency is detected alerts are sent to the rescues, also is notified by appropriate ringtone or vibration patterns in the smartphone	Indoors/outdoors	
[27]	The patient is monitored in mobility. The system can respond to emergency situations and provide different levels of attention	Indoors/outdoors	✓
[28]	If an alert of some abnormal detection or heart attack occurs, the smartphone sends data and location to cloud storage. A decision-support system notifies the emergency centers	Indoors/outdoors	✓
[29]	The cloud services generate alarms to emergency services when abnormalities are detected in patients suffering from CHF while continuous monitoring	Indoors/outdoors	✓
[30]	Fall events are reported as alarms while patient is continuously monitored in mobility	Indoors/outdoors	

response. In this case, the medical alert is not sent to any specialist, caretaker, or family members, avoiding a rapid response if a first aider was closer to the patient.

The applications have a three-layer configuration, but the internal technologies vary from one application to another. The sensor devices go from two physiological measures (EEG and EEG) to five acquired parameters (heart rate, respiration rate, ACC, temperature and environmental). In the beyond-WBSN layer, the medical information of the patient is stored, processed, and displayed in a PC located at the remote or local base station, or directly transmitted to the specialist or the patient's relatives. Despite the different technological configurations, an important parameter for outdoor scenarios is the integration of a GPS. The GPS is obligatory in the WBSN architecture, to accurately locate the person in an emergency. Another important characteristic is mobility, provided by the smartphone, which is the most commonly employed for mHealth systems.

10.5.3.2 Emergency architectures for indoors and outdoors

Table 10.8 shows the WBSN architectures for each application in Table 10.6. Actually, the mHealth system architectures for outdoors could be implemented as well for indoor scenarios. Considering that the outdoor location has more limited resources, indoor locations can integrate more technological

Table 10.7 WBSN Architectures of mHealth Applications Attending Emergencies Outdoors

Refs.	Inter-WBSN		Intra-WBSN		Beyond-WBSN		
	Sensors	COM. TECH.	Mobile Unit	COM. TECH.	Local Server	Remote Server	INFRAS. CONNEC.
[17]	Biosensors ACCs	–	Smartphone with GPS	3G	Emergency service	Location and preference servers → Internet → doctors, relatives	Traditional
[18]	HR RR ACC Temp. Environ.	Bluetooth	Smartwatch or smartphone	Not specified	Emergency service	–	Distributed services
[19]	Biomedical	Bluetooth	Smartphone with GPS	3G	Local database	Remote database	Traditional
[20]	ECG EEG	Bluetooth	Tablet with GPS	3G, 4G	–	Web services → Internet → doctor's device	Cloud services

Note: ACC, accelerometer; BP, blood pressure; COM., communication; CONNEC., connectivity; ECG, electrocardiography; EEG, electroencephalography; INFRAS., infrastructure; PR, pulse rate; SpO2, oxygen saturation; TECH., technology.

Table 10.8 WBSN Architectures of mHealth Applications Attending Emergencies Both Indoors and Outdoors

Refs.	Inter-WBSN		Intra-WBSN		Beyond-WBSN		INFRAS. CONNEC.
	Sensors	COM. TECH.	Mobile Unit	COM. TECH.	Local Server	Remote Server	
[21]	ECG	Bluetooth	Smartphone	3G or Wi-Fi	Clinical decision support	Networked PC → Internet → doctors' PDA	Traditional
[22]	ECG Temp. ACC Glucose EMG SpO2 Arduino coord.	Bluetooth	Smartphone with GPS and RFID Tag	Wi-Fi or LTE	—	Web services → healthcare management system	Internet of Things
[23]	Ambient videos	—	Ambulance with cameras, GPS, USG and vital signs	LTE	Hospital center	—	Traditional
[24]	ACC gyroscope GPS	Bluetooth	Smartphone	3G/GSM		Family and healthcare providers	Traditional
[25]	RFID tags (for user's movements) GCS BS Hearing	RFID/ Wi-Fi	APs Smartphone	Wi-Fi	Database server	Caregiver's PDA/ smartphone	Traditional
[26]	PPG	Bluetooth	Smartphone	Wi-Fi, 2G, 3G, 4G	—	Web services → doctors, users	IoT and cloud services
[27]	BP ECG ACC Temp.	Wi-Fi with P2P	Smartphone	Wi-Fi with P2P	Base station	Hospital, doctor's PC	M2M

Continued

Table 10.8 WBSN Architectures of mHealth Applications Attending Emergencies Both Indoors and Outdoors—cont'd

Refs.	Inter-WBSN		Intra-WBSN		Beyond-WBSN		
	Sensors	COM. TECH.	Mobile Unit	COM. TECH.	Local Server	Remote Server	INFRAS. CONNEC.
[28]	EEG	Bluetooth	Smartphone with microphone and GPS	UWB, RSSI on WLAN, LBS, Wi-Fi, 4G, and/or RFID	–	Web services → emergency services, doctors	Cloud services
[29]	Glucometer CO2 level SpO2 ECG	Bluetooth	Tablet or smartphone with GPS	Wi-Fi, LTE	–	Web services → MC, doctors, ambulance	IoT and cloud services
[30]	ACC	Bluetooth	Smartphone	Cellular network	–	Web services	IoT and cloud services

Note: ACC, accelerometer; AP, access point; BAN, body area network; BP, blood pressure; COM, communication; CONNEC., connectivity; coord., coordinator; ECG, electrocardiography; EEG, electroencephalography; EHR, electronic health record; EMG, electromyography; Environ., environmental; GUI, graphic user interface; HR, heart rate; implement., implementation; INFRAS, infrastructure; NDN, named data networking; PAN, personal area network; PDA, personal digital assistant; perform., performance; Phys., physiological; PPG, photoplethysmography; PR, pulse rate; RR, respiratory rate; SpO2, oxygen saturation; TECH., technology; Temp., temperature; USG, ultrasonography; VANET, vehicular ad hoc network; Videoconf., videoconference.

resources and be more suitable for network communications. A good example is a personal emergency response system with detection on heart attacks to send an alarm to the clinical decision support or specialist. As the alarms are not data demanding, the alarm could be sent over a low-speed communication network such as 3G [21]. But for indoors, a high-speed network such as Wi-Fi could be used to send all the ECG data in real time.

10.6 DESIGN OF WBSN FOR EMERGENCY PREHOSPITAL ASSISTANCE

Once a WBSN is adopted as a solution, several issues must be visualized and analyzed. On one side, we have the physician's demands: real-time communications, reliability, stable communications, and a dynamic and adaptive system to allow assisting with and tracking diverse patient conditions under different emergency scenarios; an emergency event in a suburban areas is not the same as in urban areas or at home or in outdoor locations. Furthermore, accurate design of the system and its implementation, with careful selection of each component of the WBSN, are key to obtaining the outcomes desired. During emergencies, monitoring in real time will generate a large amount of data, and a high number of attempts to transmit information from patients could cause loss of data packets and therefore a waste of resources and throughput of the WBSN.

10.6.1 WBSN ARCHITECTURE REQUIREMENTS FOR OUTDOORS

As we mentioned earlier, most of the existing WBSN architectures are based on the purpose of the project and the requirements are different for each application. However, platform design and important requirements must be satisfied for mHealth systems in outdoor scenarios. Therefore, we point out the most representative requirements that must be handled for WBSN design:

- *Sensor nodes.* The sensor nodes must satisfy full integration, with a small size. The node must be energy efficient and some techniques should be applied: local processing of multiple signals, reduction in power consumption of the microprocessor, reduction in wireless data transmission, data intensive collection, and duty cycle reduction of the sensor node.
- *Communication.* Each sensor can communicate wirelessly with other sensors or devices within WBSN (see Fig. 10.2); however, there are factors that can increase successful communications and affect the performance of the WBSN:
  - Collisions when two or more sensors try to transmit at a time (inefficient communication) causing loss of information.
  - Energy wasting due to retransmissions caused by collisions or loss of information.
  - Mobility of patient: when a patient is moving within different areas, especially in outdoor scenarios, data loss or retransmissions could occur if the WBSN is not designed with mobility support.
  - Monitoring atmosphere: heavy traffic information from external devices transmitting information, or geographic areas that prevent proper communication between WBSN devices, could interfere in the communication.

- Availability of the network: Once data is collected, the availability of the network is critical. For accurate monitoring, diagnostics, and treatment, the medical staff must be able to access patient information efficiently, at any time required.
- *Adaptability*. A WBSN must be dynamic and adaptive. Each patient will represent different health conditions, which means different monitoring priorities in terms of the energy consumption, data collection, and data transmission from each sensor.
- *Physiological signal processing*. When there is a large amount of data to transmit, signal-processing techniques gain relevance in the avoidance of energy wasting, by transmitting unprocessed data instead of processing data. For this reason, it makes sense to explore some algorithms such as QRS peak detection, QRS complex, interpolation algorithms, and compressive sensing.
- *Scheduling*. An accurate scheduling of data transmission makes the WBSN more efficient. With the physicians' support, it is necessary to study and define monitoring requirements to set the correct transfer rate of each vital sign.
- *Sampling strategies*. A constant sample rate is wasteful. It is important to use sampling strategies to reduce the volume of sample data and therefore the energy consumption caused by data transmission.
- *Quality of service*. Because physiological data is necessary for the right diagnosis and treatment, the WBSN must be always available for monitoring and vital sign transmission without errors; data must be available at the right time and in real time.
- *Patient's localization*. For patients undergoing emergency events, every single second is gold and could be the difference between survival or not. A WBSN cannot afford to waste time due to a wrong patient location.
- *Critical time parameters*. The system must be reliable. Some parameters play an important role within a WBSN. Delay, packet errors, and packet losses could affect medical diagnosis. If a vital sign measurement is lost, arrives delayed, or in a different time slot, a critical condition or sudden emergency event could be ignored. So, it is relevant to evaluate these kinds of factors when a WBSN is used.
- *Security*. Once data is created, transferred, stored, and processed, it is fundamental to set out rules/policies to ensure patient records are kept safe from intruders. Medical applications must be capable of verifying that the information transmitted is sent from a trusted source.

10.6.2 EMERGENT CONNECTIVE TECHNOLOGIES FOR WBSNS

Emergent connective technologies are introducing innovative ways of networking and cognitive data delivery [33]. The potential of connective-based architectures for enabling the specialized delivery of health-related information is introduced with technologies like cloud services, vehicular ad hoc networks (VANET), Internet of things (IoT), multiple BANS networking, and centralized and distributed services.

- *Cloud Services*. A named data networking (NDN) technology is implemented to deliver rich media content from the cloud, such as healthcare video adaptive streaming [34]. Also, the cloud services support a range of capabilities of storing, processing, and networking. Cloud data offers the possibility to scale up the system with the number of monitored patients, increasing the size of data and interconnected components, as in Ref. [20].

- *VANETs*. This type of technology offers more stable communication in the presence of high mobility, as in vehicles. The mobile network could be employed inside the ambulance to support mobile connectivity between the hospital and the ambulance in the case of critical cases [35].
- *IoT*. This novel paradigm is the pervasive ability to interconnect a variety of things that cooperate with each other. This concept enables intelligence capabilities for real-time monitoring uninterruptedly, allowing emergencies to be detected immediately. The physical devices and patients become virtual interconnected entities and can be monitored from web services [22].
- *Multiple BANS networking*. The network is formed by clusters consisting of mobile devices to distribute the data traffic until it reaches a nearby access point (AP). Only the emergency data is transmitted directly through a cellular network with a reserved channel [31].
- *Centralized and distributed services*. In this architecture the connectivity is adaptable. In centralized mode, the WBSN only connects with the healthcare station. In distributed mode, the WBSN connects with the healthcare station and sends the data to a medical display coordinator, which visualizes the information [36].

10.6.3 WBSN ARCHITECTURE ENVISIONED

First aiders could benefit from an integrated and complete solution for a WBSN platform. The technologies envisioned for the architecture of mHealth systems in outdoor scenarios, according to the application deployments in the last few years, is shown in Fig. 10.4. The fundamental devices for the WBSN system are physiological and environmental sensors, and a mobile device with GPS and video technology capabilities. In terms of communication technologies, high-speed cellular networks are needed, as well as ZigBee and Bluetooth for short range.

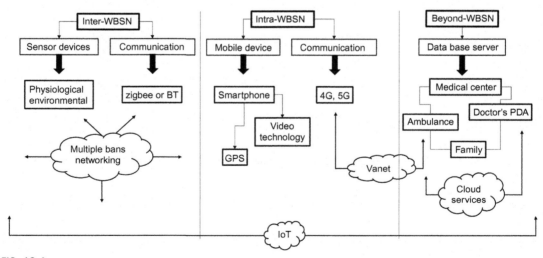

FIG. 10.4

WBSN architecture envisioned for benefiting first aiders in emergency outdoor scenarios.

It is not necessary to integrate all the emergent connective technologies depicted in Fig. 10.4. The technologies integration depends on the necessity of the context. As an example, if there are multiple injured persons, the multiple BANS networking could be integrated; if the patient is transferred to the medical center in the ambulance, the VANET is implemented. The different medical data destinations (medical center, family, doctors, etc.) could access the processed information via cloud services.

10.7 TECHNOLOGICAL ISSUES AND CURRENT BOUNDARIES

The proposed integrated and dynamic system introduces the need to overcome several technological challenges and boundaries, to continuously improve its reliability and usability required for the remote monitoring of emergencies in outdoor scenarios. These challenges and boundaries are present in the type and way the patient information is collected, the communication and service requirements, and the energy consumption of the system.

In terms of collected information, signal-processing techniques can be introduced to improve the way of collecting patient information. These techniques can compress the physiological data and reduce the volume of data to be transmitted over the WBSN. Promising signal-processing techniques are: QRS peak detection, QRS complex, interpolation algorithms, and compressive sensing. Correlating vital-sign signals is another promising technique to improve the collecting of patient information. Health issues could be advised of before an emergency event occurs, e.g., heart attack. Polling intervals could be adapted to identify correlation between specific vital signs, resulting in improved diagnosis and improved data communication performance. Sampling strategies are another possible recollection improvement. Digitizing the sensor signal at a constant sample rate is wasteful [37]. Based on previous readings and the patient criticality, an adjustable sampling strategy can reduce the volume of the sampled data as well as the energy consumption needed for data transmission [38]. Video and images are additional types of information useful to improve diagnosis and treatment. Video and images can provide the physician with valuable patient information, such as skin tone, muscle stiffness, rapid breathing, among others.

Evaluation and incorporation of new energy-efficient protocols for WBSN is one of the challenges in terms of communication requirements. For protocol selection, researchers must consider data rate and payload of protocol to improve energy efficiency [39]. Possible protocols that can extend the WBSN lifetime are: BLE, ZigBee, RFID, ANT+, NFC, and LTE technologies. Protocol selection must consider aspects such as mobility, availability at the location, and coexistence with other communication technologies in crowded environments. Strategies for improving bandwidth, avoiding data loss, and controlling latency are challenges in terms of communication and service requirements. Satisfying quality of service (QoS) needs ensures medical data without errors is available at the right time. Protecting the patient's medical information transmitted over a WBSN is another challenge in terms of communication and service requirements. Medical information can be accessed by different users, including medical staff, the patient's family, and insurance companies; however, the privileges of accessing and handling those sensitive data should not be the same. To control and regulate these issues, a high level of policies is required [40].

For the benefit of medical data protection, the Health Insurance Portability and Accountability Act (HIPAA) has established some regulations for different kinds of medical services [41]. Adopting strategies to reduce network traffic due to frequent patient's data updating is one of the challenges in terms of energy optimization. It has been reported that, in emergency situations, PHI must be upgraded every 10 seconds [2]. This frequent updating of patient data represents a large amount of data to be transmitted over the network. A possible strategy is to categorize sensors with a critical-condition priority such as heart attack, stroke, respiratory arrest, etc. Another strategy is the adoption of scheduling techniques to avoid data loss because of several sensors trying to access the channel at the same time. Avoiding system failure due to discharged batteries is yet another challenge in terms of energy optimization. The system fails if one or more of the components runs out of power. WBSNs need to be designed and implemented in terms of energy efficiency, and there are three main challenges to face: reducing microprocessor power consumption of components, reducing wireless data transmission, and decreasing time of node receiver active mode [42]. In addition, it is necessary to develop and incorporate power-aware policies capable of adapting to the monitoring requirements, determined by the network traffic and the patient [43].

10.8 CONCLUSIONS

The lack of an mHealth integrated platform solution means that first aiders in emergency situations do not have all the available technologies to give proper assistance to the person in need. Assuming such a system's development, as previously depicted, some advantages are envisioned.

First, caregivers could increase their resources for accurate tracking of an emergency; physicians at the hospital could carry out real-time monitoring of the scene and the patient's medical condition. Real-time monitoring means an enormous improvement as compared to the current system, where the communication between physicians and first aiders is limited to two or three times, which translates into slower response to patients facing crisis during their transfer to a hospital. Moreover, first aiders could avoid unnecessary transfers to hospital when a remote evaluation of a patient's status reveals that the patient's life is not in danger and only preventive measures are needed; an example of this occurs when a patient suffering a chronic degenerative disease, such as high-blood pressure, calls emergency services because he is feeling dizzy and has a severe headache. Most of the patients in this scenario would be anxious and probably would demand to be assisted at a hospital, because the patient would feel more comfortable after receiving specialized medical attention. MHealth systems could offer this kind of attention at the patient's location, thus avoiding expensive ambulance fees.

Another specific case which could benefit from an mHealth system is when a patient suffering a heart attack needs to be treated with thrombolysis or angioplasty methods [44]. The difference between living or dying with sequels is related to the timely application of these methods, and an mHealth system based on physicians advising first aiders could allow their precise utilization in a timely manner. Most of the mHealth system platforms have traditional connectivity, and the medical data of the patient is transmitted hierarchically. Gradually, the implementation of emergent connective technologies is redefining information connectivity and is replacing the traditional WBSN systems.

Finally, there are many challenges needing to be overcome by researchers in the field; the most important were highlighted in the previous section, but mHealth education training for prehospital personnel is also required. There is no doubt that telemedicine in general will continue in its constant evolution, due to its role in future and present medical practices; however, at the present time, we are still in the process of development and acceptance within the medical community. We can confidently predict that, once mHealth systems meet the requirements of operating as adaptive, dynamic, and smart platforms, death rates related to emergency situations will be considerably reduced.

REFERENCES

[1] Kirtava Z, Gegenava T, Gegenava M. In: mHealth for cardiac patients telemonitoring and integrated care. In proc. of IEEE 15th international conference on e-health networking applications & Services, 21–25, October; 2013.

[2] Gonzalez E, Peña R, Vargas-Rosales C, Avila A, de Cerio DP-D. Survey of WBSNs for pre-hospital assistance: trends to maximize the network lifetime and video transmission techniques. Sensors 2015;15:11993–2021.

[3] OMS, Estadísticas sanitarias mundiales, Organización mundial de la salud (World Health Organization); 2012. Available from, http://www.who.int/gho/publications/world_health_statistics/ES_WHS2012_Full. pdf. [accessed 16 August 2016].

[4] Bowles KH, Dykes P, Demiris G. The use of health information technology to improve care and outcomes for older adults. Res Gerontol Nurs 2015;8:5–10.

[5] La fundación RACC. La "hora de oro", 60 minutos vitales tras un accidente 2011, Available from, http://consejosconducir.racc.es/es/-la-hora-de-oro–60-minutos-vitales-tras-un-accidente; 2012. [accessed 3 December 2012].

[5a] Xu J, Murphy SL, Kochanek KD, Bastian BA. Deaths: final data for 2013. Natl Vital Stat Rep 2016;64(2):1–118.

[6] Goldberg C. A practical guide to clinical medicine. San Diego: University of California; 2015. Available from, https://meded.ucsd.edu/clinicalmed/vital.htm. [accessed 27 December 2016].

[7] O'Neill D, Le Grove A. Monitoring ill patients in accident and emergency. Nurs Times 2003;99:32–5.

[8] Aquino-Santos R, González-Potes A, Rangel-Licea V, García-Ruiz MA, Villaseñor-González LA, Edwards-Block A. Wireless communication protocol based on EDF for wireless body sensor networks. J Appl Res Technol 2008;6:120–8.

[9] Sasser S, Varghese M, Kellermann A, Lormand JD. Prehospital trauma care systems. Geneva: World Health Organization; 2005. Available from, http://www.who.int/violence_injury_prevention/publications/services/ 39162_oms_new.pdf. [accessed 14 January 2017].

[10] Curso de Soporte Básico de Vida (CSBV). Programa USAID/OFDA/LAC de capacitación y asistencia técnica. In: Oficina de asistencia para desastres del gobierno de los estados Unidos de Norte América (OFDA). 2009 Available from, https://scms.usaid.gov/sites/default/files/documents/1866/MR%20-%20SBV.pdf. [accessed 18 January 2017].

[11] EMS/Hospital Radio Communications. http://www.health.ri.gov/publications/guides/EMSHospitalRadio CommunicationsFieldOperations.pdf; 2017. [accessed 21 January 2017].

[12] Ramírez AJL, Ocampo LR, Pérez PI, Velázquez TD, Yarza SME. Importance of effective communication as a factor of quality and safeness in medical attention. Acta Med Austriaca 2011;9:167–74.

[13] Riccardo C, Flavia M, Ramona R, Chiara B, Roberto V. A survey on wireless body area networks: technologies and design challenges. IEEE Commun Surv Tutorials 2014;16(3):1635–57.

[14] Hao Y, Foster R. Wireless body sensor networks for health-monitoring applications. Physiol Meas 2008;29(11):R27–56.

[15] Baig MM, Gholam-Hosseini H, Connolly MJ. Mobile healthcare applications: system design review, critical issues and challenges. Australas Phys Eng Sci Med 2015;38(1):23–38.

[16] Yoglín R, Richard W, Nelem P, Azzedine B. Monitoring patients via a secure and mobile healthcare system. IEEE Wirel Commun 2010;17(2):59–65.

[17] Solanas A, Martinez-Balleste A, Perez-Martinez PA, Fernandez de la Pena A, Ramos J. m-Carer: privacy-aware monitoring for people with mild cognitive impairment and dementia. IEEE J Sel Areas Commun 2013;31(9):19–27.

[18] Castillejo P, Martínez JF, López L, Gregorio R. An internet of things approach for managing smart services provided by wearable devices. Int J Distrib Sens Netw 2013;9(2):1–9.

[19] Banos O, Villalonga C, Garcia R, Saez A, Damas M, Holgado-Terriza JA, et al. Design, implementation and validation of a novel open framework for agile development of mobile health applications. Biomed Eng Online 2015;14(Suppl. 2):S6.

[20] Serhani MA, Menshawy ME, Benharref A. SME2EM: smart mobile end-to-end monitoring architecture for life-long diseases. Comput Biol Med 2016;68:137–54.

[21] Bansal A, Kumar S, Bajpai A, Tiwari VN, Nayak M, Venkatesan S, et al. Remote health monitoring system for detecting cardiac disorders. IET Syst Biol 2015;9(6):309–14.

[22] Hussain A, Wenbi R, Lopes da Silva A, Nadher M, Mudhish M. Health and emergency-care platform for the elderly and disabled people in the smart city. J Syst Softw 2015;110:253–63.

[23] Cicalò S, Mazzotti M, Moretti S, Tralli V, Chiani M. Multiple video delivery in m-health emergency applications. IEEE Trans Multimedia 2016;18(10):1988–2001.

[24] He J, Hu C, Wang X. A smart device enabled system for autonomous fall detection and alert. Int J Distrib Sens Netw 2016;12(2):1–10.

[25] Kbar G, Al-Daraiseh A, Mian HS, Abidi MH. Utilizing sensors networks to develop a smart and context-aware solution for people with disabilities at the workplace (design and implementation). Int J Distrib Sens Netw 2016;12(9):1–25.

[26] Bellagente P, Crema C, Depari A, Ferrari P, Flammini A, Rinaldi S, et al. The "Smartstone": using smartphones as a telehealth gateway for senior citizens. IFAC-Papers OnLine 2016;49(30):221–6.

[27] Chung K, Kim JC, Park RC. Knowledge-based health service considering user convenience using hybrid Wi-Fi p2p. Inf Technol Manag 2016;17(1):67–80.

[28] Shamim-Hossain M. Cloud-supported cyber-physical localization framework for patients monitoring. IEEE Syst J 2017;11(1):118–27.

[29] Abawajy JH, Mehedi-Hassan M. Federated internet of things and cloud computing pervasive patient health monitoring system. IEEE Commun Mag 2017;55(1):48–53.

[30] Gravina R, Ma C, Pace P, Aloi G, Russo W, Li W, et al. Cloud-based Activity-aaService cyber-physical framework for human activity monitoring in mobility. Futur Gener Comput Syst 2017;75:158–71.

[31] Kim Y, Lee S. Energy-efficient wireless hospital sensor networking for remote patient monitoring. Inf Sci 2014;282:332–49.

[32] Chen C, Knoll A, Wichmann H-E, Horsch A. A review of three-layer wireless body sensor network systems in healthcare for continuous monitoring. J Mod Intern Things 2013;2(3):24–34.

[33] Doukas C, Fotiou N, Polyzos GC, Maglogiannis I. Cognitive and context-aware assistive environments using future internet technologies. Univ Access Inf Soc 2014;13(1):59–72.

[34] Chen M. NDNC-BAN: supporting rich media healthcare services via named data networking in cloud-assisted wireless body area networks. Inf Sci 2014;284:142–56.

[35] Alghamdi B, Fouchal H. A mobile wireless body area network platform. J Comput Sci 2014;5(4):664–74.

[36] Khan Z, Aslam N, Sivakumar S, Phillips W. Energy-aware peering routing protocol for indoor hospital body area network communication. Procedia Comput Sci 2012;10:188–96.

[37] Alippi C, Anastasi G, Francesco M, Roveri M. Energy management in wireless sensor networks with energy-hungry sensors. IEEE Instrum Meas Mag 2009;12(2):16–23.

[38] Salim C, Makhoul A, Darazi R, Couturier R. In: Adaptive sampling algorithms with local emergency detection for energy saving in wireless body sensor networks. NOMS 2016–2016 IEEE/IFIP network operations and management symposium; 2016. p. 745–9.

[39] Mahmoud M, Mohamad A. A study of efficient power consumption wireless communication techniques/modules for internet of things (IoT) applications. Adv Intern Things 2016;6:19–29.

[40] Al-Janabi S, Al-Shourbaji I, Shojafar M, Shamshirband S. Survey of main challenges (security and privacy) in wireless body area networks for healthcare applications. Egypt Inform J 2017;18:113–22.

[41] Hash J, Bowen P, Johnson A, Smith CD, Steinberg DI. An introductory resource guide for implementing the Health Insurance Portability and Accountability Act (HIPAA) security rule. Gaithersburg, Maryland, USA: National Institute of Standard and Technology, 2005.

[42] Celic L. Energy efficiency in wireless sensor networks for healthcare. Available from. https://www.fer.unizg. hr/_download/repository/KDI,_Luka_Celic.pdf; 2012. [accessed 25 September 2016].

[43] Quwaider M, Jararweh Y. A cloud supported model for efficient community health awareness. Pervasive Mob Comput 2016;28:35–50.

[44] Johnston S, Brightwell R, Ziman M. Paramedics and pre-hospital management of acute myocardial infarction: diagnosis and reperfusion. Emerg Med J 2006;23(5):331–4.

FURTHER READING

[45] Xu J, Murphy SL, Kochanek KD, Bastian BA. Deaths: final data for 2013. Natl Vital Stat Rep 2016;64(2):1–119.

INDOOR ACTIVITY TRACKING FOR ELDERLY USING INTELLIGENT SENSORS

Nelson Wai-Hung Tsang*, Kam-Yiu Lam*, Umair M. Qureshi*, Joseph Kee-Yin Ng[†], Ioannis Papavasileiou[‡], Song Han[‡]

City University of Hong Kong, Kowloon Tong, Hong Kong* Baptist University of Hong Kong, Kowloon Tong, Hong Kong[†] University of Connecticut, Storrs, CT, United States[‡]

11.1 INTRODUCTION

An aging population is a global trend of modern society. As people live longer, more and more elderly will stay at home alone and need to handle daily living activities by themselves, for example, preparing meals, watching TV, and reading newspapers. In order to minimize the risk of having home accidents, it is important to measure the fitness of their self-caring abilities by tracking their daily activities [1]. In this chapter, we discuss how the latest intelligent sensor technologies can be applied for tracking the common indoor activities performed by an elderly person during the day. Our target users are those elderly who can perform simple living activity, for example, preparing simple meals and going to the toilet by themselves, but have problems in movement and need to stay alone within their living quarters most of the time during the day. Therefore, they may need a cane or walker to help them move, and their movement within their living quarters is limited.

We discuss how different types of sensor devices, for example, 3D depth cameras and motion sensors, can be used for detecting the activities performed during a day [2]. Based on tracking the daily activities of an elderly person, we can generate various *activity measures* to describe his/her current health status and self-catering abilities [3]. For example, in addition to knowing how much time the user spends sleeping in the afternoon, which normally is a poor health indicator, we may also want to know the time that he/she spends watching TV and reading the newspaper during the day, as this is considered to be a good health indicator. A rapid decrease in good activity measure values will raise an alert on the current health status of the user.

Another important objective of tracking the indoor activity of the user is to handle emergency situations. If the user has an accident at home (e.g., falling on the floor), it is important to generate an alert message to the user's relatives so that prompt help can be provided [4]. Measuring the speed of step movement can also be an important indicator on the probability of having falls, as well as an important health measure [5]. Normally, the step movement speed of the elderly decreases with age. It is important to encourage them to have more daily activities and exercises in order to maintain health and muscle strength.

Intelligent Data Sensing and Processing for Health and Well-being Applications. https://doi.org/10.1016/B978-0-12-812130-6.00011-1

In this chapter, we first discuss how Kinect, a 3D depth camera, can be used for activity detection through the introduction of our system, called *SmartMind*. An important benefit of using Kinect for activity detection is that it does not require the users to carry any sensors. This can minimize the impact of the system operation on their daily life. Furthermore, the 3D images of the user captured by Kinect provide a lot of information useful in estimating the current activities of the user as well as generating various activity measures. However, privacy concerns may arise, since many users will not like to be continuously monitored by a camera. Therefore, in the second part of the chapter, to resolve the limitations of *SmartMind*, we introduce a reduced version of *SmartMind*, called *ActiveLife*, which is designed and developed using simple motion sensors (e.g., accelerometers, gyroscopes, and magnetometers [6]). The main benefits of using motion sensors are their low cost and simple set-up in addition to having no privacy concerns. Since most of the motion sensors are small, for example, smaller than a match box, they are very convenient to be carried by the users. However, improving the accuracy of activity estimation using motion sensors is a challenging issue. In the final part of the chapter, we show the effectiveness of using machine learning methods, such as support vector machines (SVMs), in improving the accuracy of activity estimation in *ActiveLife*.

11.2 ACTIVITY DETECTION AND TRACKING USING KINECT

SmartMind is an indoor activity-tracking system for the elderly. The sensor adopted in *SmartMind* is Kinect v2.0 [7], which is a 3D depth camera developed by Microsoft that it captures images and generates 3D posture data of the user continuously. The captured 3D posture data are forwarded to a server for further analysis and decision making, for example, activity detection and fall detection.

11.2.1 THE KINECT DEVICE

Fig. 11.1 shows the sensor specification of a Kinect v2.0 device, which has several different types of sensors. It includes a color sensor (a camera) to capture the RGB channel data in 1280×960 pixel resolution of the color image. Furthermore, it has an infrared (IR) emitter and an IR depth sensor that emits and reads the IR beams, respectively, for estimating the depth of the main body joints of the captured images. The captured body joint data are converted into depth information for measuring the distance between the user and the sensor device (the Kinect device). Each generated frame of image streams captured by the Kinect consists of color, depth, and body joint data that are captured at the same time point for analysis. In addition, the Kinect device contains a multiarray microphone to record audio for finding the location of the sound source and the direction of the audio wave. A three-axis accelerometer is also configured to determine the orientation of the device.

11.2.2 LIVING ROOMS AND ACTIVITIES

Fig. 11.2 shows a typical living environment for an elderly person in Hong Kong. Normally, the living room is small with simple furniture. For example, it has a three-seat sofa facing a long bench with drawers, and a TV set is placed on top of the long bench. On the other side of the living room is a small dining table with one or two chairs. Next to the bedroom is the toilet. On the other side of the toilet is a small kitchen for preparing meals.

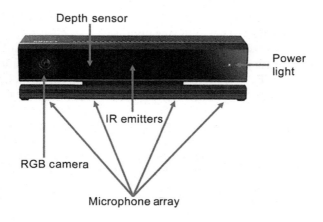

FIG. 11.1

The sensors of Kinect v2.0.

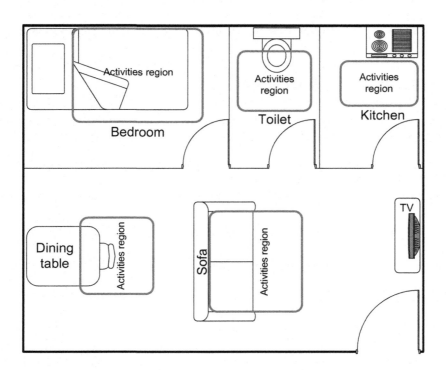

FIG. 11.2

An example floor plan of the living room and activity regions.

It is common that most of the daily activities are performed in the living room, in particular near the sofa and the dining table. Therefore, in the following discussions, we concentrate on the activities performed in the living room. These activities can be classified into three basic postures: Standing, Lying, and Sitting. They are called basic postures since they are the easiest to be identified by Kinect (to be discussed in the next section). To simplify the activity detection process, we adopt a *context-based approach* in defining the activities [8]. In the context-based approach, we divide the living environment into different activity regions, as shown in Fig. 11.2, and the basic postures are context-constrained according to the activity regions, that is, a basic posture is conditioned with one or several activity regions. For example, standing can only be performed in the open regions, which refers to the regions without any furniture items, for example, sofa and dining table. Lying may take place on the sofa, in the bed, or in open regions, but not on the dining table. If lying occurs in an open region, it may be identified as a "fall" or "doing exercises." Note that the constraints are defined according to the common habits of the user to be tracked and the setting of the living environment.

Table 11.1 summarizes the set of activities under different basic postures according to the activity regions. For example, under the basic posture of "sitting" are listed "having meal" and "watching TV." Under the basic posture of "Standing," we see "moving in the open region," "moving into bedroom," "moving out of bedroom," "moving into kitchen," "moving out of kitchen," "moving out of bathroom," "moving into bathroom," and "doing exercises." Note that it is not our purpose to track all the activities performed by the user during a day, which is too difficult and not necessary. Instead, our objective is to obtain information on important activities to describe the current health status of the user. For example, the user may move from the sofa to the bench to get something and then return to the sofa. This activity is not common and is not our interest.

11.2.3 OVERVIEW OF THE SMARTMIND SYSTEM

Fig. 11.3 gives an overview of the *SmartMind* system and summarizes its main components. These consist of the *SmartMind Server*, the *SmartDB Server*, the *SmartWeb Server*, the Kinect devices, a group of databases, and an "app" called *SmartReminder* running on the smartphone to receive

Table 11.1 The Set of Activities to Be Tracked

Basic Posture	Location	Activities
Lying	Sofa	Sleeping in sofa
	Open area	Fall on floor
		Doing exercises
	Bed	Sleeping
Sitting	Sofa	Watching TV
		Doing exercises
	Dining table	Having meal
		Reading newspaper
Standing	Open area	Moving into or out of bedroom
		Moving into or out of bathroom
		Moving into or out of kitchen
		Doing exercises

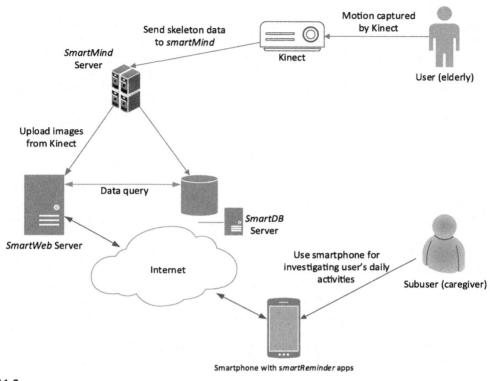

FIG. 11.3

The system architecture of *SmartMind*.

notification messages and browse information obtained from the *SmartWeb* server. Depending on the size and the settings of the living room, two or more Kinect devices are installed at fixed locations in the living room. For example, to cover the whole living room, one of them may be placed on top of the dining table and another one may be put above the TV set. Note that in order to capture the motions of the user correctly, the user must be at least 1.5 m from the Kinect device [7]. Otherwise, the generated posture data may contain a large number of errors. Note that here we concentrate on tracking the activities of the user within the living room. Further tracking functions can easily be performed in other functional rooms by installing additional Kinect devices pointing to the inside of the functional rooms. If the functional room is very small and the posture data cannot be generated correctly by the Kinect device, we may use motion sensors to capture the movements of the user. How motion sensors can be used for activity detection is discussed in Section 11.3.

SmartMind Server is implemented in C#. It connects to the Kinect devices through a wireless network (e.g., Wi-Fi). On receiving images and 3D posture data from the Kinect devices, *SmartMind Server* performs analysis to determine the current activity of the user. If the detected activity is an "Important Activity" predefined by the system administrator, an image of the activity will be taken by the Kinect device. Then, an activity record including a simple description together with the image of the activity will be stored into the database maintained by *SmartDB Server* as an activity log. The administrator may be a relative of the target user who knows the living habits and abilities and possible risks of the target user. Some of the important activities to be monitored by *SmartMind* are critical ones, such as "falling on the floor accidentally." If such an activity is identified, the associated emergency action maintained in the *emergency database* will be triggered. For example, a message together with the image of the scene will be sent immediately to the smartphones of the relatives or caregivers maintained in the *contact database*. *SmartDB Server* is an SQL server. It maintains a group of databases, for example, activity database, emergency database, contact database, and room layout database. The room layout database describes the layout of the rooms and system settings. The activity database records the activities of the user within the living room. *SmartWeb Server* runs on IIS [9]. It provides an Internet interface for reviewing the activity records maintained by *SmartDB Server*.

11.2.4 ACTIVITY DETECTION AND TRACKING

Various efficient techniques have been proposed for motion capture and activity detection [2, 3, 10, 11]. In this section, we show how Kinect can be used for activity detection.

11.2.4.1 Posture captured by Kinect

To identify the current activity of the user, we need both: (1) current posture and (2) current activity region. For a user being tracked, each Kinect device creates a skeleton (called posture data), which consists of 20 positioned points. Each point is called a "joint" of the user, as shown in Fig. 11.4 [7]. Each joint is represented in Cartesian coordinates to express its 3D location, as shown in Fig. 11.5. The location of the Kinect device is set to be the origin, that is, $x = 0$, $y = 0$, and $z = 0$ (measured in meters), and a negative value of x represents the location on the left-hand side of the Kinect device.

By checking the coordinates of different joints obtained from a Kinect device, we can estimate the current posture of the user. For example, Fig. 11.6A–C shows three example postures: "Standing," "Sitting," and "Lying." They are easy to identify since the conditions for the involved joints have significant differences. Fig. 11.6A shows that the user is "Sitting" and the difference in vertical height between the knees and hip center is small (e.g., less than 25 cm) and the angle formed by the hip center, knee, and ankle is greater than 90 degrees (e.g., around 100 degrees). As shown in Fig. 11.6B, for a "Standing" posture, the knee joint of the user is at a position well below the hip joint while the angle formed by the head, hip, and knee is usually more than 150 degrees. If the head and hip joints are at similar levels and the angle formed by the head, hip center, and knee is also large (e.g., greater than 150 degrees), the posture of the user is "Lying" (Fig. 11.6C).

The posture data to be used for posture identification include: (1) head, (2) spine shoulder (shoulder center), (3) spine base (hip center), (4) left/right hand, (5) left/right shoulder, (6) left/right elbow, (7) left/right foot, (8) left/right knee, (9) spine mid, and (10) left/right ankle. The wrists and feet are very close to hands and ankles, respectively. Therefore, we neglect them in identifying the posture. In total, we have 16 effective joints for each image frame obtained from a Kinect device. Each of these vectors consists of a 3D representation of the joints at each instance (e.g., x-coordinate, y-coordinate, and z-coordinate). All the coordinates are measured from the hip joint as the origin. After obtaining the joint data from Kinect, the *SmartMind Server* performs a preprocessing to transform the joint data of each frame from Cartesian values to polar coordinates. As shown in Fig. 11.7, the polar representation

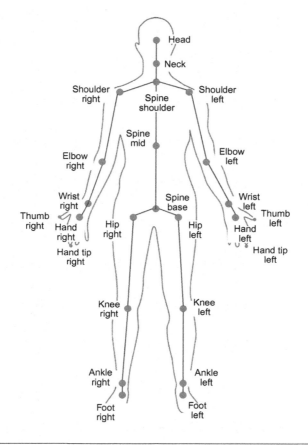

FIG. 11.4

Joints in the Skeleton for Kinect v2.0 [7].

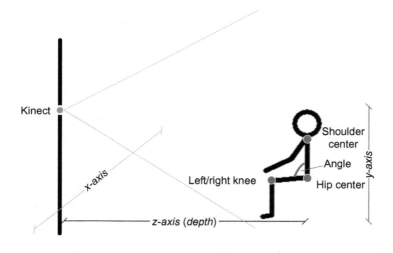

FIG. 11.5

Coordinates of the joints of the user.

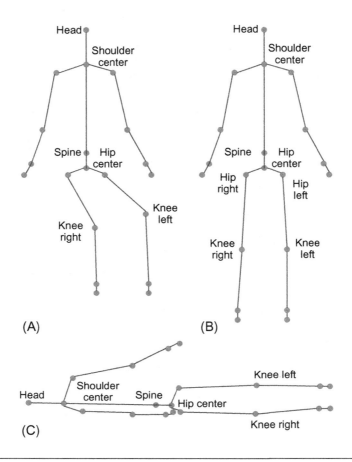

FIG. 11.6

Basic postures: (A) Sitting posture; (B) standing posture; (C) lying posture.

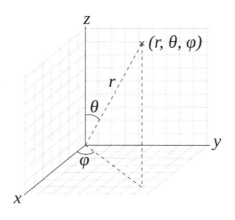

FIG. 11.7

Polar coordinates.

of a point consists of its distance r from the origin (i.e., the hip joint), the azimuthal angle φ (angle measured from the x axis), and the polar angle θ (measured from the z axis). After generating a new set of posture data, we compare the data with the values of the previous set of posture data for checking of posture. If their values are similar to those of the previous posture data, that is, most of the differences in values are within a threshold value, we may conclude that the user remains in the same posture and continues the same activity. Otherwise, we will determine the new posture of the user, and then determine the new activity after obtaining the current activity region. Note that the generation of the joints of the target user may be affected by other objects. If it is a stationary object, for example, a table, we can easily ignore the occluded joints and only concentrate on the detected joints for posture detection. If the object is a moving object, for example, a friend of the target user is visiting and the visitor's movement has occluded most of the joints of the target user, we may skip the set of sampled joints for posture detection until the number of detected joints is large enough.

If more than one activity has the same posture and is located in the same activity region, we need further checking on the relative locations of the body joints. For example, watching TV, sitting for rest, and reading the newspaper all have the same posture of sitting and the same activity region of the sofa. For this case, we need to check the relative locations of the arms of the user. If the relative locations of the arms are separated and maintained at a height similar to the shoulders, for example, greater than 0.5 m, continuously for a period of more than 1 min, we may conclude that the user is reading the newspaper. Otherwise, the user is watching TV, if the Kinect detects that there are audio sounds from the TV set. If there is no sound from the TV set and the arms are significantly lower than the shoulders and remain at similar locations, this indicates that the user is sitting at rest.

11.2.4.2 Identification of location
Obtaining the current activity region of the user within his or her living room is accomplished through the depth camera embedded in the Kinect device and the layout of the living room. Each furniture item in the living room (e.g., sofa, dining table, etc.) or other functional room is defined as an activity region by two pairs of x and z (e.g., a rectangle formed by $<x_1,z_1>$ and $<x_2,z_2>$) as shown in Fig. 11.2. From a Kinect device, the posture data of the user is generated and the coordinates (i.e., x and z) of each joint are extracted. Note that here, we concentrate on the x-coordinate and z-coordinate. We do not need to check the y-coordinate which is the height of the joint. If the coordinates of the main joints (i.e., head, hip center, and spine) of the user are within an activity region, for example, sofa, the user is considered to be inside that activity region.

In addition to identifying the current activity region of the user, we can use the transition between activity regions to measure the step speed of the user. This is an important measure on the probability of falling. If the average step speed is very slow, this indicates that the self-caring ability of the user is poor and a movement problem may exist. Furthermore, based on the activity region information, we can determine which region is the user's favorite due to spending most of the time in that region.

11.2.4.3 Activity transition
Some posture data obtained from Kinect may contain errors. The problem becomes more serious if the user is too close to the Kinect device (e.g., closer than 1.5 m). Therefore it is important to design an efficient mechanism to filter out the errors and concentrate on detecting important activities. Otherwise, the *flip-flop problem* may occur, that is, the activities of the user change quickly between two or three activities due to the errors in the captured posture data from the Kinect. To resolve the

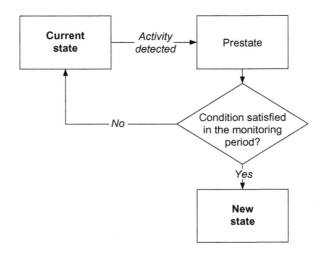

FIG. 11.8

State transition of activities.

flip-flop problem, we use a state transition to identify a new activity. As shown in Fig. 11.8, on receiving a set of posture data from the Kinect, we search for a prestate. We first identify the basic posture (e.g., lying, sitting, or standing) to be performed by the user. For example, the user is at the dining table and the basic posture is "sitting." Then, the set of activities under the basic posture of "sitting" at the dining table will be searched to determine the prestate of the user. If the user's hands are moving toward the mouth, the user will be determined to be "having meal" as the prestate. Then, the "having meal" prestate will be checked continuously for a predefined period of time (e.g., 30 s in this case). If the same prestate occurs repeatedly during the monitoring period, the prestate is confirmed and the user's state transits to "having meal at the dining table." After recording the start time of an activity, we will look for the end time of the activity by repeating the same procedure to identify when the user enters into another activity (state). Note that the duration of the period for confirming a prestate to be a new state is activity dependent and is a predefined system parameter.

11.2.4.4 Fall detection

A fall is a highly critical activity and its detection needs to be very accurate. Various efficient fall-detection algorithms have been proposed for different types of sensors [4, 11–14]. Most of the techniques involve carrying accelerometers and gyroscopes for detecting the speed of movement of the user. In the following, we show how the posture data obtained from the Kinect can be used for fall detection. One of the important observations from our preliminary studies is that the step movements of elderly people are normally slow. They may walk slowly within their living rooms and some of them may even require a cane or walker to assist them in walking. Therefore, the main approach adopted in fall detection in *SmartMind* is the speed or acceleration of movement, especially if the user is moving within a small area vertically (i.e., falling down). By comparing the vertical length along the y-coordinate of all the joints (*dH*) with the horizontal length of the x-coordinate of the joint (*dW*), we can easily determine whether the user is lying down or not. For example, the ratio of *dH:dW* should be greater than 1 in standing and sitting postures. However, the ratio will be less than 1 if the posture is lying. The lying posture may be because the user is doing exercises on the floor. Therefore, in order to increase the accuracy of fall detection, we add two more conditions; the vertical position of the head ($Head(Y)$) must be lower than 0.5 m, and the vertical movement speed of the head must be faster than a threshold that is greater than the maximum movement speed of the user (e.g., 2 m/s).

Fig. 11.9 summarizes the conditions for fall detection. As shown in the figure, it requires a fast drop in height of the main body joints of the user (i.e., the head) to indicate a fall for a distance of more than

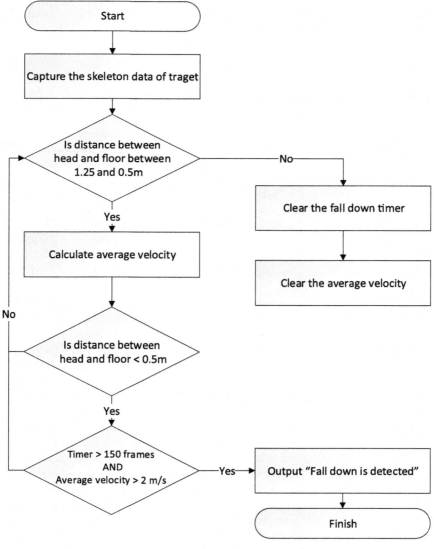

FIG. 11.9

Conditions for fall detection.

half the height of the user. From the captured posture data of the user, if the distance between the center of the shoulder of the user and the floor is smaller than 1.25 m and larger than 0.5 m, we calculate the falling speed of the user. If the average speed of this motion is greater than a certain value (e.g., 2 m/s) and the user (i.e., the hip joint) remains close to the floor (e.g., less than 0.5 m) for more than a predefined period of time (e.g., 5 s), a "Fall" is assumed. Then, the scene will be captured by the Kinect device and sent together with a notification to the registered contact people immediately. Note that requiring the user to remain on the floor is an important condition in fall detection. If the user can regain a standing or sitting posture after the "fall," this implies that the "fall" is probably not a serious one and the problem can be handled by the user. At any rate, the system records the fall so that the number of falls that occurs during the day can be reported to the relatives and/or caregivers.

11.3 ACTIVITY DETECTION AND TRACKING USING MOTION SENSORS

Although Kinect is an effective sensor for tracking indoor activities, it has privacy concerns, as most users do not like to be continuously monitored through video. To resolve this limitation of Kinect, we have designed a reduced version of *SmartMind*, called *ActiveLife*, to illustrate how simple motion sensors, for example, accelerometers, gyroscopes, and magnetometers [6], can be adopted for activity tracking. The main benefits of *ActiveLife* as compared with *SmartMind* are its low cost and simple set-up. In *ActiveLife*, we do not need to install any fixed devices for activity detection. Instead, we only require the user to carry a small motion sensor package, which is usually much cheaper in cost, on the body. Although this will raise the problem of the user forgetting to wear the motion sensors, this problem can easily be resolved by generating alarms if no motion is detected for a period of time (e.g., 30 min) to remind the user to put on the sensor device. Furthermore, with the rapid advances in electronic technologies, various tiny sensors are now being produced and some of them can even be embedded into our clothing [15].

The key challenge in the design of *ActiveLife* is how to maintain accuracy in activity detection. The motion sensors only report the physical sensor values, such as changes in angular and movement velocities, and they lack any location information of the user within the living environment. In addition, we do not want the user to carry too many sensors attached at different parts of the body to track the body posture, since this will make the set-up very complex and could be quite inconvenient to the user. Therefore, how to characterize the sensor data for different activities is the main problem that we need to handle in the use of motion sensors for activity detection. As demonstrated in *SmartMind*, being aware of the current activity region is very important for detecting the current activity. Of course, the location of the user can be provided by various indoor localization techniques (e.g., using Wi-Fi and Bluetooth). We would like to minimize the complexity and cost of the system set-up in the design of *ActiveLife*. We do not include any positioning techniques for localization in *ActiveLife*. Instead, we model the system environment using the characteristics of the generated sensor data values. In order to simplify the activity detection process, we also apply a *context-based approach* as adopted in *SmartMind* to model the activities and the transitions between different activities within the living rooms. It is assumed that both the furniture in the living room and the settings of the functional rooms are fixed. With the context-based approach, when the motion sensors give sensor data unrelated to the context, the sensor data will be ignored until they can be identified to have a particular context.

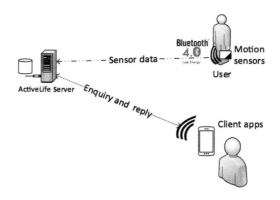

FIG. 11.10

The system components of *ActiveLife*.

11.3.1 OVERVIEW OF THE ACTIVELIFE SYSTEM

Fig. 11.10 gives an overview of *ActiveLife* and summarizes its main system components. It consists of a motion sensor package (including multiple motion sensors) carried by the user, an *ActiveLife Server* and a smartphone as a client. In our current prototype development, we use the *SensorTag* [16], which contains 10 different types of sensors, including an accelerometer, a gyroscope, and a magnetometer. Each of them is a triaxial sensor and reports measured values in three axes (e.g., *x*-axis, *y*-axis, and *z*-axis). The generated sensor data values are transmitted to the *Activelife Server* through low-energy Bluetooth (BLE), which provides a communication range of around 30 m. This is sufficient to cover most of the living environments of the of elderly. The received sensor data is then be converted to higher level activity data and used for generating activity statistics. This information is stored into a local database maintained by the *ActiveLife Server*. Similar to *SmartMind*, the stored activity data and statistics can be retrieved through a smartphone app.

Similar to *SmartMind*, we divide the living rooms into activity regions and different activity regions have their predefined activities. For example, cooking is performed in the kitchen and sleeping is performed in the bedroom. Therefore, once we have determined the current activity to be performed by the user, we can also know the current location region of the user. If more than one activity can be performed in an activity region, in order to distinguish them, we need to further check the posture of the user or even the movement characteristics of the user. We assume that the activity regions are fixed (i.e., the furniture in the rooms is fixed). Therefore, the trajectory of moving from one region to another region is more or less fixed (i.e., the movement distance between two activity regions is similar and the user usually chooses the shortest path).

Similar to *SmartMind*, we classify the set of activities to be tracked by *ActiveLife* into three *basic postures*: *standing*, *sitting*, and *lying*. Furthermore, each activity is associated with a predefined orientation, called *heading direction*, in the activity region. For example, when the user is watching TV, the heading direction of the user is toward the TV set. The heading direction of the user is another important piece of information that we use to help estimate the current activity region. Different activities have their own heading directions according to the settings of the living rooms.

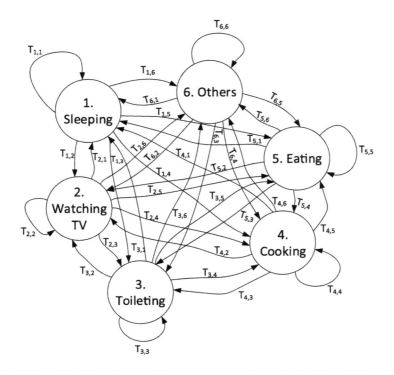

FIG. 11.11

The state diagram of the activity transitions.

Fig. 11.11 shows the transitions of the set of defined activities to be performed by the user within the living rooms. Note that each state is characterized with a heading direction and a contextual attribute called *movement distance* of the transition. Since we assume that the settings of the living rooms are fixed, the movement distance of the same transition is the same while the movement distances of different transitions are different. Therefore, after knowing the current activity, we can determine the next activity by checking the movement distance of the transition and the heading direction of the new activity.

11.3.2 GENERATION OF SENSOR DATA

In *ActiveLife*, we adopt the *MPU-9250* 9-axis MEMS motion sensor (i.e., accelerometer, gyroscope, and magnetometer) in *SensorTag* [16] to compute the basic posture, movement distance, and the heading direction of the user for each activity and transition. The gyroscope measures the changes in angular acceleration while the accelerometer measures the changes in linear acceleration. The magnetometer measures the magnetic field strength. Each of the motion sensors generates three streams of sensor data for three axes (i.e., x-axis, y-axis, and z-axis), according to the defined frequency. For example, let us say that the sensor is placed horizontally on a table (i.e., the z-axis is perpendicular to the earth while the

x-axis and y-axis are parallel to the earth and perpendicular to each other). Since the z-axis of the accelerometer is perpendicular to the earth when it is being placed on the table, it experiences the gravity acceleration of a value close to $g = -9.8\text{m/s}^2$ if the sensor remains stationary. Since the raw data reported by a magnetometer is the strength of magnetic field, a better way to use the data for analysis is to convert the magnetic field data to direction as the heading direction of the user when he or she is carrying the sensors. Similar to the conversion of posture data in *SmartMind*, as shown in Fig. 11.7, we can determine the azimuth of the user by using the following equation:

$$Azimuth = arcTan(y/x) \tag{11.1}$$

Fig. 11.12 shows the algorithm to calculate the azimuth of the user using the reading from the x-axis and y-axis of the magnetometer [17]. Note that the calculation of azimuth can only be applied when the user is not engaged in high gravity motion (e.g., walking).

An important concern in the design of *ActiveLife* is where to put the motion sensors. In our settings, we place the *SensorTag* on the thigh of one of the legs so that the x-axis of the motion sensors is always perpendicular to the gravity when the user is standing, while the y-axis is perpendicular to the gravity when he or she is sitting. The gravity force from x-axis to y-axis can help to distinguish between the postures of standing and sitting. This information will help us to determine his or her current activity as shown in the following text and figures. Furthermore, since the location of the motion sensor is fixed, we can use the magnetometer to determine the heading direction of the user, for example, the direction in which the knee of the user is facing.

To identify the important features of the sensor data for different postures and transitions, an experiment was performed using *SensorTag* which was set to report sensor data every 25 ms. The movements of the tester were summarized in Fig. 11.13. First, the tester stood at P_1 for 5 s and then walked toward P_2 at normal speed. When he reached P_2, he stood for another 10 s and then changed his heading direction and ran to P_3. After arriving at P_3, he changed his heading direction and sat down for 10 s. Afterwards, he stood up and walked back to P_1, and then laid down for 10 s before the end of the experiment. Note that in this experiment, we use the postures (e.g., lying, standing, and sitting) to represent the activities, while walking and running are transitions.

```
Input: X, Y
Output: Azimuth
IF X = 0 { IF Y < 0
                Azimuth = 90;
          ELSE
                Azimuth = 270;}
ELSE {
        IF X < 0
              Azimuth = 180 - arcTan(Y/X) * 180/π;
        ELSE
              IF Y < 0
                    Azimuth = - arcTan(Y/X) * 180/π;
              ELSE
                    Azimuth = 360 - arcTan(Y/X) * 180/π;
```

FIG. 11.12

The algorithm for calculating azimuth.

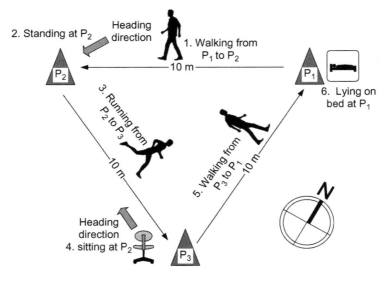

FIG. 11.13

Experimental setup.

The experiments were repeated three times, and Fig. 11.14A–C summarizes the sensor data readings collected from the accelerometer, gyroscope, and magnetometer, respectively, in one of the experiments. It is obvious to see in Fig. 11.14A (accelerometer data) and B (gyroscope data) that the sensor data obtained for transitions (walking and running) are very different from those obtained from the activities (standing, sitting, and lying). The sensor data of the accelerometer and gyroscope for walking and running show periodic patterns while the sensor data during the standstill periods (e.g., activities) are mostly close to zero, indicating no or only a few motions of the legs. Similarly, the activity periods can roughly be determined by the direction obtained from the magnetometer (Fig. 11.14C). The high variation of direction values during walking and running is due to the high errors in calculation of azimuth values under high gravity motion. Therefore, the sensor data obtained from the magnetometer will only be used during the standstill periods.

The sensor data from the accelerometer consist of two components: gravity and acceleration, due to the movement of the leg. To remove the gravity component, we apply a filter [18] and the values of the input parameters are summarized in Table 11.2. Fig. 11.14D shows the filtered amplitudes calculated using the following equation:

$$Amp_{Acc} = \sqrt{(A_x)^2 + (A_y)^2 + (A_z)^2} \tag{11.2}$$

The filtered amplitudes clearly show the differences in values between transitions and activities. To identify the frequency of walking and running motions, we apply the discrete fast Fourier transform

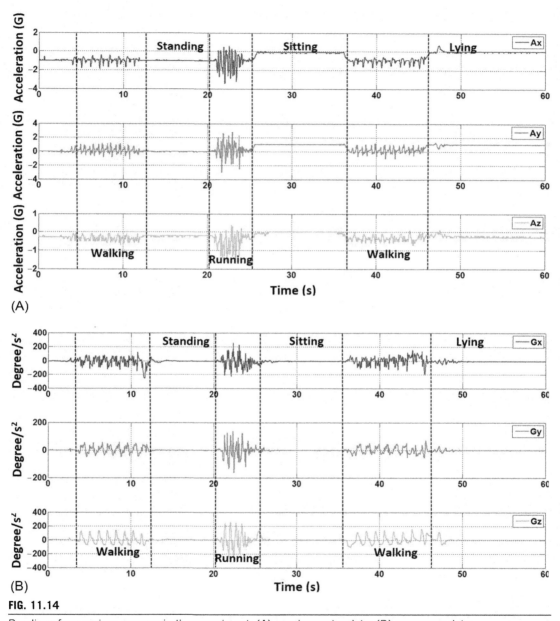

FIG. 11.14

Readings from various sensors in the experiment: (A) accelerometer data; (B) gyroscope data;

(Continued)

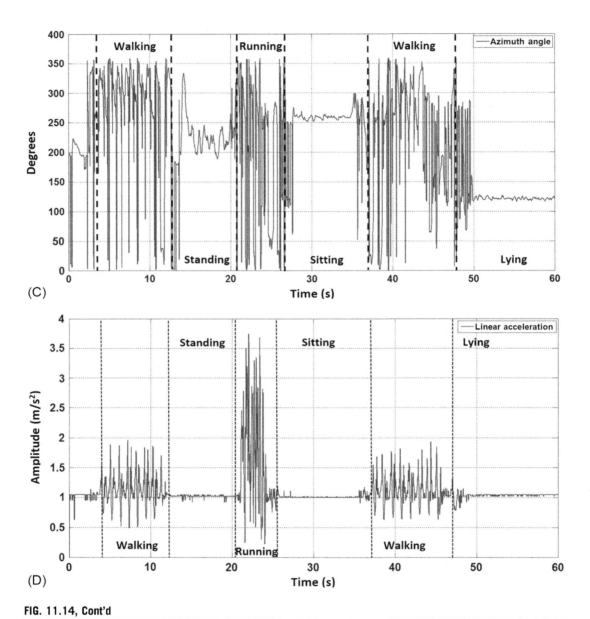

FIG. 11.14, Cont'd

(C) heading directions of user; (D) combined acceleration amplitude.

Table 11.2 Filter Design Parameters
Design Parameters and Design Specifications
Sampling frequency = 40 Hz
Passband frequency = 0.8 Hz
Stopband frequency = 0.4 Hz
Passband amplitude = 1 dB
Stopband amplitude = 80 dB

(FFT) [19] to transform each acceleration data sample (n) to a frequency domain represented by frequency bins (f) using the following equation:

$$Mag_{Acc}(f) = \sum_{n=0}^{N-1} Amp_{Acc}(n) * e^{\frac{-2\pi jn(f)}{N}},$$ (11.3)

where N is the total number of samples.

Table 11.3 summarizes the distribution of the accelerometer sensor data for different activities (postures) and transitions (walking and running). The periodic amplitudes of the sensor data correspond to the periodic movements of the legs when the tester was walking or running. Please note that the standard deviations of walking and running are much higher than those of activities (lying, sitting, and standing) due to the greater movement of the leg. By using these features of the sensor data obtained from the accelerometer and gyroscope, we can easily distinguish between transitions and activities. Since running is moving faster than walking, the maximum amplitude for running is also higher than that of walking. In addition to higher amplitude, the period is also shorter for running due to faster movement. The period can be interpreted as the period for completing one movement cycle (i.e., two steps), since we only put the SensorTag on one of the legs of the tester. The number of periods in a transition indicates the number of steps performed during the transition. This can be used as a measure of the movement distance of the transition to distinguish different transitions, as it is assumed that the distance of each step is the same. Different transitions have different movement distances and can be represented by the number of steps performed.

Table 11.3 Activity and Transition Features					
Activity	**Max (m/s²)**	**Min (m/s²)**	**Mean**	**Standard Deviation**	**Frequency (Hz)**
Standing	1.049	0.922	1.021	0.022	–
Sitting	1.183	0.78	1.013	0.029	–
Lying	1.049	0.922	1.038	0.012	–
Walking	1.934	0.635	1.123	0.254	10
Running	3.75	0.223	1.512	0.797	8

Although standing, sitting, and lying all show low values from the accelerometer, their values are not exactly the same. If we examine Fig. 11.14A carefully, we can see that, while standing, the sensor data values for the y-axis and z-axis are mostly at zero while those for the x-axis are around $-1g$ (one gravity acceleration). This is because in standing posture, the y-axis and z-axis are parallel to gravity while the x-axis is perpendicular to gravity. However, when the posture is sitting or lying, the y-axis is perpendicular to gravity. Therefore, we see that the sensor data values for the y-axis are now around $1g$ while those for the x-axis and z-axis are close to zero. If we carefully examine the sensor data values during the lying periods, we can also see that, although the data values of the x- and z-axes are similar to those of sitting, the z-axis data have values slightly smaller than those of the x-axis, as we normally bend our leg slightly toward the ground while lying. Therefore, the basic postures of sitting and standing can be distinguished according to the sensor data for the x-axis and y-axis of the accelerometer. For the lying posture, the sensor data from the y-axis of the accelerometer will change from zero to one and the sensor data for the z-axis will be slightly smaller than that of x-axis, which has values close to zero.

11.3.3 DETECTION OF ACTIVITIES

Fig. 11.15 shows the algorithm for determining activities according to the data streams obtained from the motion sensors. It consists of two steps. First, we examine the data streams from the accelerometer and gyroscope to classify them into *activity* or *transition* periods using the SVMs, which are machine learning methods. We apply SVM to achieve better accuracy in activity detection, and the details will be discussed in the next section. Different users may have different heights and walking characteristics. Therefore, different threshold values may need to be used for different users in activity classification. The benefit of using SVM is that it can learn from the training data sets to adaptively determine the threshold values for activity classification of different users.

In our algorithm, transition periods represent walking movement, while nontransition periods represent activities. After identifying the periods of the data stream for an activity, we use SVM again to classify the basic posture of the activity. In the last step, we check the heading direction of the user and the movement distance of the transition to determine the current activity.

11.3.4 SUPPORT VECTOR MACHINES

The SVM [20] is one of the most popular machine learning algorithms. It has been widely used for solving computer vision problems. SVM maps the points in a high dimensional space using a kernel. A hyperplane is used to divide the geometric space into two parts for classification. The main advantage of SVM is that it solves a convex problem and is suitable for classification of continuous features. The regularization parameters can be tweaked to control overfitting.

Assume a data set composed of labeled patterns $D_N = \{(x_i, y_i), i = 1, ..., N\}$, where the training pattern $x_i \in \mathbb{R}_n$ belongs to either of the two classes: $y_i \in +1, -1$. SVM is intrinsically a binary classifier that constructs a linear separating hyperplane to classify data instances [21–23]. The classification capabilities of traditional SVMs can be substantially enhanced through the transformation of the

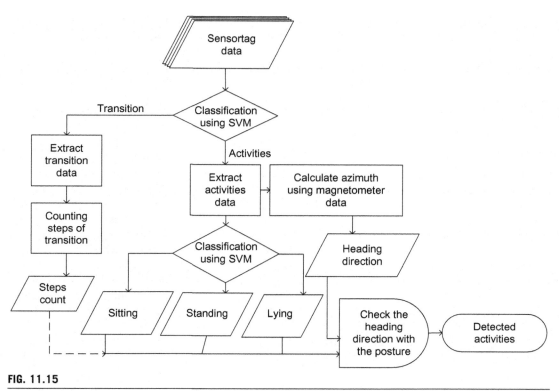

FIG. 11.15

The algorithm for classifying activities.

original feature space. Considering a nonlinear mapping $\Phi: \mathbb{R}^n \rightarrow F$, each vector x_i in the original feature space is mapped into a potentially higher dimensional feature space $F: x_i \rightarrow \Phi(x_i)$, where $i = 1, ..., N$. The SVM classifier aims at finding the optimal separating hyperplane in the transformed space F, which is an objective formulated as a quadratic optimization problem:

$$\max_{\alpha_i} \sum_{i=1}^{N} \alpha_i - \frac{1}{2} \sum_{i,j=1}^{N} \alpha_i \alpha_j y_i y_j K(x_i, x_j)$$

$$\text{subject to} \sum_{i=1}^{N} \alpha_i y_i = 0, \quad 0 \le \alpha_i \le C; \quad i = 1, ..., N \tag{11.4}$$

where α_i, with $i = 1, ..., N$, denotes the Lagrange multipliers, and C determines a tradeoff between the maximum margin and classification error. The resulting decision surface is given by

$$f(x) = \sum_{i=1}^{N} \alpha_i y_i K(x_i, x) + b. \tag{11.5}$$

In f, many of the a_i's are zero and the ones that are nonzero define the support vectors x_i that are used to calculate the decision surface. A new pattern x can be classified by calculating $sgn(f(x))$.

The kernel function K allows scalar products to be implicitly computed in F without explicitly using the mapping function Φ. Any function that satisfies the Mercer's theorem can be used as a scalar product and hence it serves as a kernel function. The kernel function used in the experiments was the radial basis function (RBF) kernel:

$$K(x_i, x_j) = \exp\left(-\gamma \| x_i - x_j \|^2\right). \tag{11.6}$$

In our model, we have applied the parameters in the SVM classification shown in Table 11.4.

Following the algorithm shown in Fig. 11.15, we performed two sets of the training model for testing the efficiency of using SVM for classification. The first one classifies whether the user is in transition (i.e., labeled "1") or not in transition (i.e., labeled "2") by using the sensor data obtained from the accelerometer. Fig. 11.16A shows the input sensor data values of the accelerometer and Fig. 11.16B shows the results of the classified transition and activity periods. After we have classified the time periods for activities, that is, the legs of the user are mostly stationary during the periods, we extract the data during each activity period and perform another classification to identify the postures (e.g., labeled "1" for standing, "2" for sitting, and "3" for lying) of the user for each period. The classification results of different postures are shown in Fig. 11.17. As shown in Figs. 11.16 and 11.17, we can have a very high accuracy (99.8%) in classifying transitions and activities, and the accuracy in classifying different postures is 100%.

Table 11.4 SVM Parameters
For Movement Classification
Kernel: RBF
$C = 100$
Epsilon $= 0.1$
Gamma $= 0.01$
For Posture Classification
Kernel: RBF
$C = 1$
Epsilon $= 0.1$
Gamma $= 0.1$

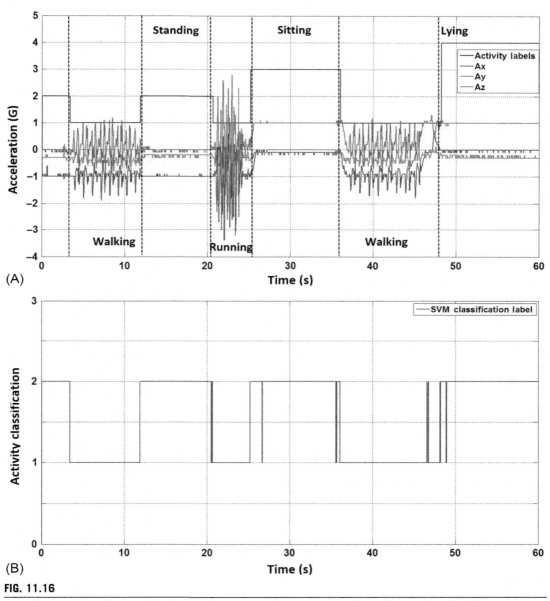

FIG. 11.16

Classification of transitions and activities using SVM: (A) variation in amplitude of accelerometer; (B) classification of transition and activities.

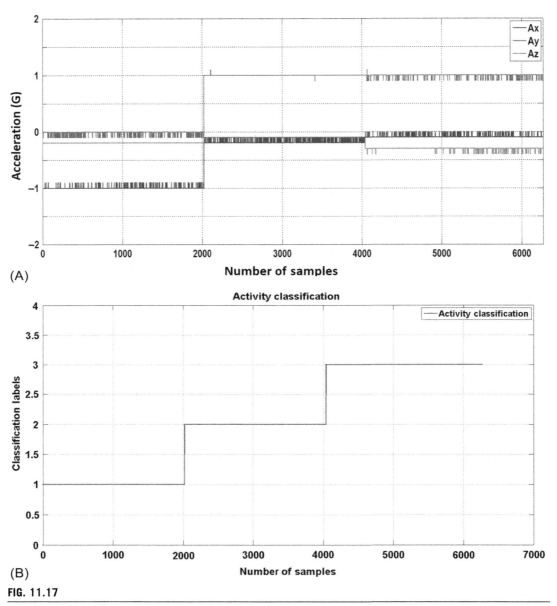

FIG. 11.17

Classification of different postures using SVM: (A) variation in amplitude of accelerometer; (B) classification of posture.

11.4 CONCLUSIONS AND FUTURE WORK

In this chapter, we discussed how to use intelligent sensor technologies for tracking the indoor daily activities of elderly persons in their living areas. We show how Kinect and the 3D depth camera can be adopted for activity detection by identifying the current posture and locations of the user. Since Kinect may have some privacy concerns, we also discussed how to use motion sensors for activity estimation, and how to improve the detection accuracy through machine learning methods. We illustrated how we adopted simple sensors, such as accelerometers, gyroscopes, and magnetometers, for monitoring daily activities. By classifying the different postures of the user, we defined the activities undertaken using a context-based approach.

For future work, we will design a device containing all the sensors that is more user-friendly to the elderly. More real tests will then be performed with our target elderly within their living apartments. This will include testing with elderly with different mobility problems and living habits. Another important future work is to extend *ActiveLife* to include a localization algorithm using Bluetooth to improve its activity detection accuracy for the cases where the set of activities may be more complicated, for example, where a greater number of different activities are performed at the same location region.

REFERENCES

[1] Dawadi PN, Cook DJ, Schmitter-Edgecombe M. Automated cognitive health assessment using smart home monitoring of complex tasks. IEEE Trans Syst Man Cybern Syst 2013;43(6):1302–13.

[2] Fosty B, Crispim-Junior CF, Badie J, Bremond F, Thonnat M. Event recognition system for older people monitoring using RGB-D camera. Workshop on assistance and service robotics in a human environment, Tokyo, Japan; 2013.

[3] Cottone P, Maida B, Morana M. User activity recognition via Kinect in an ambient intelligence scenario. In: International conference on applied computing, computer science and computer engineering; 2014. p. 49–54.

[4] Rantz MJ, Banerjee TS, Cattoor E, Scott SD, Skubic M, Popescu M. Automated fall detection with quality improvement rewind to reduce falls in hospital. J Gerontol Nurs 2014;40(1):13–7.

[5] Yodpijit N, Tavichaiyuth N, Jongprasithporn M, Songwongamarit C, Sittiwanchai T. The use of smartphone for gait analysis. In: 2017 3rd international conference on control, automation and robotics (ICCAR); 2017. p. 543–6. https://doi.org/10.1109/ICCAR.2017.7942756.

[6] Lane ND, Miluzzo E, Lu H, Peebles D, Choudhury T, Campbell AT. A survey of mobile phone sensing. IEEE Commun Mag 2010;48(9):140–50.

[7] Microsoft Crop. Kinect for windows; 2014, http://www.microsoft.com/en-us/kinectforwindows/.

[8] Worrall S, Agamennoni G, Nieto J, Nebot E. A context-based approach to vehicle behavior prediction. IEEE Intell Transp Syst Mag 2012;4(3):32–44. https://doi.org/10.1109/MITS.2012.2203230.

[9] Microsoft Crop. Iis web server overview; 2014, https://www.iis.net/learn/get-started/introduction-to-iis/iis-web-server-overview.

[10] Urturi Breton ZSD, Zapirain BG, Zorrila AM. KiMentia: Kinect based tool to help cognitive stimulation for individuals with dementia. In: IEEE international conference on e-health networking, applications and services; 2012. p. 325–8.

[11] Bian ZP, Hou J, Chau LP, Magnenat-Thalmann N. Fall detection based on body part tracking using a depth camera. IEEE J Biomed Health Inform 2015;19(2):430–9.

[12] Rougier C, Meunier J, St-Arnaud A, Rousseau J. Robust video surveillance for fall detection based on human shape deformation. IEEE Trans Circuit Theory 2011;21(5):611–22. https://doi.org/10.1109/TCSVT.2011.2129370.

[13] Chaccour K, Darazi R, Hassani AHE, Andrès E. From fall detection to fall prevention: a generic classification of fall-related systems. IEEE Sensors J 2017;17(3):812–22. https://doi.org/10.1109/JSEN.2016.2628099.

[14] Wang Y, Wu K, Ni LM. WiFall: device-free fall detection by wireless networks. IEEE Trans Mobile Comput 2017;16(2):581–94. https://doi.org/10.1109/TMC.2016.2557792.

[15] Sung M, Jeong K, Cho G. Suggestion for optimal location of textile-based ECG electrodes on an elastic shirts considering clothing pressure of the shirt. 2008 12th IEEE international symposium on wearable computers; 2008. p. 121–2. https://doi.org/10.1109/ISWC.2008.4911603.

[16] Texas Instruments. SensorTag2015 wiki. Available from: http://processors.wiki.ti.com/index.php/SensorTag2015.

[17] STMicroelectronics. Using LSM303DLH for a tilt compensated electronic compass; 2010.

[18] Williams AB, Taylors FJ. Electronic filter design handbook. New York: McGraw-Hill; 1988.

[19] Van Loan C. Computational frameworks for the fast Fourier transform. Philadelphia: Society for Industrial and Applied Mathematics; 1992, https://doi.org/10.1137/1.9781611970999.

[20] Cortes C, Vapnik V. Support-vector networks. Mach Learn 1995;20(3):273–97. https://doi.org/10.1007/BF00994018.

[21] Burges CJC. A tutorial on support vector machines for pattern recognition. Data Min Knowl Discov 1998;2(2):121–67.

[22] Muller KR, Mika S, Ratsch G, Tsuda K, Scholkopf B. An introduction to kernel-based learning algorithms. IEEE Trans Neural Netw 2001;12(2):181–201.

[23] Moustakidis SP, Theocharis JB, Giakas G. Subject recognition based on ground reaction force measurements of gait signals. IEEE Trans Syst Man Cybern B 2008;38(6):1476–85.

USER-CENTERED DATA MINING TOOL FOR SURVIVAL-MORTALITY CLASSIFICATION OF BREAST CANCER IN MEXICAN-ORIGIN WOMEN

12

Guillermo Molero-Castillo*,†, **Everardo Bárcenas***,†, **Gabriela Sánchez***,†, **Aldair Antonio-Aquino***

CONACYT, Mexico City, Mexico* *University of Veracruz, Xalapa, Mexico*†

12.1 INTRODUCTION

Currently, due to the growth of data collection and the evolution of computing power, information from many different sources is stored. This allows preservation of useful historical data to explain the past, to understand the present, and to predict future situations. Therefore, there is an increasing need to seek new ways of analyzing and processing existing data sources to obtain useful information and knowledge. However, the increasing data size stored in these sources is often a limiting factor for analysis, so that specialized technologies have been developed to process and obtain information of interest with the purpose of serving as support in the decision-making process [1,2]. One of these technologies is data mining, which allows a smart data analysis, and solves two major challenges [3]: (a) extraction and discovery of knowledge, and (b) analysis and identification of trends and behaviors. This smart data analysis allows a better understanding of the phenomena implicitly described, and helps in the decision-making process.

Information obtained through data mining can be valuable when using appropriate methods and tools, so it is now necessary to define methodologies to plan and guide the development of projects related to data analysis. Currently, there are several methodologies, such as CRISP-DM, KDD, SEMMA, Catalyst, Six Sigma, among others, with different approaches and operations, which are formed by stages that help structure a data mining project in a formal process of development [4]. However, these methodologies, despite their wide variety, often do not consider users as an important factor in each of their stages, thus limiting user participation in each of the stages they comprise. As a consequence, data mining developments have limitations of usability and accessibility and are even lacking in functionality [5,6]. There is then a natural need for new methodologies that consider users during each of the stages.

Intelligent Data Sensing and Processing for Health and Well-being Applications. https://doi.org/10.1016/B978-0-12-812130-6.00012-3

Involving users gives a greater value to human factors, collaborative work, and the incremental and systematic development of data mining projects. For instance, in current data mining methodologies, user satisfaction is not considered in evaluation steps. The user-centered approach proposed in the current work encompasses the usability principle as an important factor for improving user satisfaction and accessibility [7]. In addition, not only particular users are considered, but all those users involved in the project are taken into consideration, considering the ages, capacities, traits, differences, and other characteristics of interest.

The importance of users is fundamental in projects of knowledge discovery through data analysis, because they all have different knowledge, preferences, cognitive styles, and other mental skills. A user analysis is required in order to obtain a better understanding of needs [6,8]. Furthermore, the importance of users in data mining projects is not only in the exploration of data volumes for knowledge discovery, but also in the decision-making process through the use of interactive tools. These tools must be easy to use, learn, and remember [9]. In addition, in order to involve users as part of the data mining process, the following features must be considered [10]: (a) privacy, to preserve the identity of the users; (b) customization, to ensure that the user benefits from the knowledge found; (c) portability, to ensure that the data flow is everywhere; and (d) proficiency, to provide sufficient resources for knowledge discovery.

Users are an essential part of data mining projects, so it is then crucial to involve them from the analysis and understanding of the project to the validation and results presentation. It is important to note that an incorrect interpretation of user needs and requirements could lead to the failure of projects could or limit user expectations [7]. This chapter presents research results on a user-centered data mining methodology. This methodology is based on the principles of the ISO 9241-210:2010 standard (design of user-centered interactive systems). Our proposal is applied to the analysis of clinical data over the survival and mortality of Mexican-origin women diagnosed with breast cancer. The data source is composed of the database of the Surveillance, Epidemiology, and End Results (SEER) program of the National Cancer Institute (NCI) of the United States.

12.2 BACKGROUND
12.2.1 DATA MINING METHODOLOGIES

There are a wide variety of data mining tasks and techniques, so a number of methodologies or processes for the development of projects are appearing. Table 12.1 provides a summary of the main features of the current state of the art of data mining methodologies, which in the last decades have increased significantly.

One of the best-known data mining processes is knowledge discovery in database (KDD), which consists of a series of sequentially structured stages for knowledge generation and decision making. Among the main features of this process is its iterative nature and its orientation to user actions [11]; however, this process does not describe the specific tasks and activities that must be performed in each process stage. Specifics about these tasks and activities are then subject to the development team [12].

Another data mining methodology is SEMMA (Sample, Explore, Modify, Model, Assess), created by the SAS Institute (Statistical Analysis Systems), which defines SEMMA as the selection, exploration, and modeling of large volumes of data to discover patterns of interest, such as: to segment customer groups, to identify profitable customers, and those that go with the competition, among other

Table 12.1 Data Mining Methodologies: Main Features

	KDD	CRISP-DM	SEMMA	Catalyst	Six Sigma
Phases	– Integration and collection – Selection, cleansing and transformation – Data mining – Evaluation and interpretation – Diffusion	– Business understanding – Data understanding – Data preparation – Modeling – Evaluation – Deployment	– Sampling – Exploration – Modification – Modeling – Evaluation	– Data preparation – Modeling – Refine the model – Implement the model – Communication of results	– Definition – Measurement – Analysis – Improvement – Control
Iterative steps	Yes	Yes	No	Yes	No
Tools	Free and commercial	Free and commercial	Commercial	Free and commercial	Free and commercial
Evaluation	Results evaluation based on project objectives	Results evaluation based on model performance and project objectives	Results evaluation based on model performance	Results evaluation based on project objectives	Results evaluation based on model performance
Creation year	1996	1999	1998	2003	1986

tasks [13]. Particularly, SEMMA begins with an exploratory analysis of data, leaving aside the analysis and prior understanding of the project [14]. SEMMA, like the KDD process, does not describe the specific activities that must be performed in each process stage. In addition, SEMMA is particularly related to the use of commercial products of the SAS Institute.

CRISP-DM (Cross Industry Standard Process for Data Mining) is another process currently used in data mining projects [12,15]. This process is characterized by dividing the project into different phases, tasks, and activities [16]. These phases follow an iterative process, useful for making changes in previous stages. However, it does not include control tasks and monitoring of the work plan.

Another methodology is Catalyst, known as P3TQ (Product, Place, Price, Time, Quantity), composed of two models [17,18]: (a) business (M-II), and (b) information exploitation (M-III). M-II offers a guide for the development of (business opportunity) problem models, and M-III provides a design and execution guide of data mining models.

An industrial process adapted to data mining is Six Sigma, defined as an organized and systematic method for process improvement, new products, and services based on statistical and scientific methods in order to reduce error rates or failures [19]. Six Sigma involves the analysis of data through statistical tools to reduce variation through continuous improvement.

Although these methodologies meet the main objective of guiding the discovery of patterns of interest in data volumes, they still lack important aspects, such as major user participation, considering all the different roles in each stage, and the efficient presentation of the patterns obtained. These aspects are fundamental for a better explanation and understanding of the new knowledge obtained. In this sense, interest arises in developing user-centered data mining with the purpose of improving user satisfaction and experience in the data mining process.

12.2.2 BREAST CANCER

Breast cancer is a malignant tumor that originates in the cells of the breast. These cells grow in a disorderly manner and invade surrounding tissues, as well as distant organs [20]. This disordered growth occurs when new cells are produced when the body does not need them, and old cells do not die when they should die [21]. These cells that are not necessary for the body could eventually form a mass of tissue, known as a cancerous tumor [22]. Factors such as migration, new life habits, and the aging of populations make breast cancer a major threat in various parts of the world, currently representing one of the three main causes of female death in Latin America [22,23].

The possibility of cure and improvement in the quality of life of patients with breast cancer depends on the extent of the disease at the time of diagnosis and the adequate application of available knowledge and resources, increasing efficiency and the technical quality being used for this scientific evidence [24]. In this sense, there is increasing interest in carrying out research from the scientific and technological points of view, to develop new support tools to identify disease behaviors and trends.

12.2.2.1 Symptoms and risk factors

Women with breast cancer may experience different symptoms, which are not definite and which may be due to other health problems. However, in the case of presenting some symptoms, it is advisable to go to a specialist in order to have a diagnosis for possible early treatment. Some of the most common symptoms are [25,26]: change in size or shape of the breast; nipple plunged inward; injuries to the nipple area; pain in the breast, particularly pain that does not go away; thickening of the breast, near the breast, or in the armpit; skin irritation or changes in the skin; and secretion, especially if blood is present.

There are certain risk factors that could increase the risk of breast cancer. Some known risk factors are [25,27]: (a) *age*, as breast cancer risk increases with respect to age, most cases originating from 50 years old; (b) *personal history*, as cancer on one breast significantly increases risk in the other breast; (c) *family history*, as breast cancer in the family, especially mother, sister or daughter, increases risk, even more so when a family member had cancer before 40 years old; (d) *lifestyle*, as lack of physical activity, alcohol consumption, and poor diet are other risk factors; (e) *ovarian cancer*, as risks of breast cancer increase from previous ovarian cancer; (f) *radiation*, as exposure to ionizing radiation at an early age may increase the risk of developing breast cancer; (g) *race*, as white women are more likely to develop this disease compared to black or Asian women (although in general breast cancer can affect all women regardless of race and ethnicity); (h) *oral contraceptives or birth control pills*, as some studies point to the likelihood of developing breast cancer due to contraceptive use; and (i) *breast density*, as women with larger areas of dense tissue are at increased risk for breast cancer (a mammogram, or X-ray of the breast, shows whether the breast tissue is dense or adipose).

12.2.2.2 Treatment

Tumors of breast cancer are classified according to their size and degree of development. Several treatments are defined according to the classification of tumors. According to NCI [28] there are three main treatments:

- Surgery. It is the main treatment for breast cancer in the early stages. These cancers can be treated by surgery with breast conservation, also known as lumpectomy or partial mastectomy. There is also radical mastectomy, which consists of breast excision.

- Radiotherapy. It is a common treatment in breast cancer patients, done to reduce the chance that cancer will return after surgery (breast-conserving or mastectomy). Treatment depends on tumor size and medical evaluation.
- Chemotherapy (systemic therapy). This is the use of anticancer drugs or drugs that help fight the growth of cancer cells. These medications can be applied at different stages of the disease.

12.3 **METHODOLOGY**

User involvement in data mining projects, in their different roles as data and information providers, business specialists, decision makers, data scientists, data and systems analysts, software architects, software engineers, among others, in addition to implying a better acquisition of requirements, also results in more customized projects [29]. Further, end users involved in the project must be considered, regarding their respective ages, capacities, features, and other characteristics [30]. One of the disciplines studying the acquisition of user needs and incorporating them into final products is user-centered design (UCD), which is widely used for the design and development of technological projects [31].

It is important to note that an incorrect interpretation of user needs and requirements could lead to the failure of data mining projects, or could limit user expectations. Thus, adopting user-centered methods in diverse solutions favors improvements in efficiency, effectiveness, well-being, and user satisfaction [7]. In this work, ISO 9241–210:2010 is included as a user-centered design method. This standard of the International Organization for Standardization (ISO) provides principles that serve as a reference to ensure a user-centered process in the design of interactive systems. It also includes the fundamentals of CRISP-DM as a data mining process. Recent studies highlight that CRISP-DM has a greater acceptance, with 43%, compared to other methodologies, such as SEMMA (8.5%) and KDD (7.5%). These studies also showed a high acceptance of own or customized methodologies (that is, tailor-made) with 27.5% acceptance [15].

Both processes, ISO 9241-210:2010 and CRISP-DM, provide significant foundations that jointly define stages and concrete actions to carry out user-centered data mining projects. Fig. 12.1 describes a general projection of user-centered data mining divided into the following phases: (a) contextual analysis, which encompasses the understanding and description of stakeholders, and also defines the objectives of the project and of data mining, and an overall project plan; (b) data analysis, which consists of an approximation of the understanding of the data; (c) data preparation, used to obtain the minable data view on which the techniques of data mining are applied; (d) modeling, which contemplates the selection of one or more data mining techniques to find useful data patterns, trends, or new knowledge, depending on the needs of the project and users (use of free or commercial tools or new custom applications); (e) evaluation, which consists of the evaluation of results from the point of view of the project and user objectives; and (f) presentation, which includes the presentation of the results obtained through interactive interfaces.

In the methodology, the objective is to involve the user in significant stages of data mining, following an iterative cycle in order to obtain requirements, needs, activities, work environments, and other aspects of interest related to the development of the project. We describe in Fig. 12.2 the stages and activities of the proposed user-centered data mining methodology, highlighting the user-centered design.

User involvement in the methodology is given by different roles: (a) data and information providers (stages: contextual analysis, data analysis, data preparation) provide data sources and information for

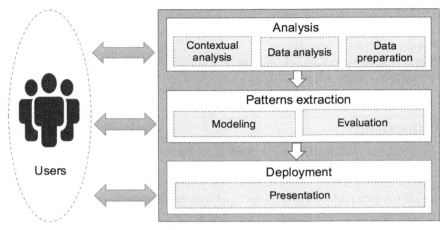

FIG. 12.1

General structure of the user-centered data mining methodology.

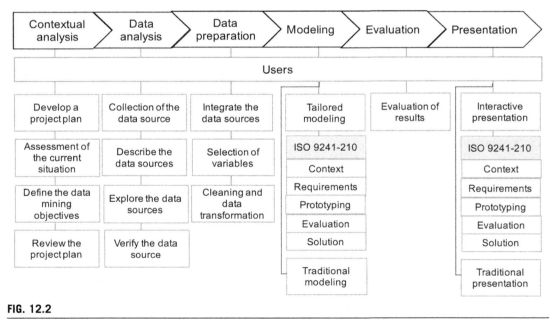

FIG. 12.2

Stages and activities of the user-centered data mining methodology.

the understanding of project objectives and the data; (b) business specialists (stages: contextual analysis, data analysis, data preparation, modeling, evaluation, presentation) are responsible for providing and validating information about the understanding of project objectives, requirements, and needs of end users; (c) decision makers (stages: contextual analysis, data analysis, presentation) can be specialists or those in charge of activity management in the organization; and (d) the development team, consisting of data scientists, data and systems analysts, software architects, software engineers, user experience specialists, visualization specialists, test engineers, among others (stages: contextual analysis, data analysis, data preparation, modeling, evaluation, presentation), design and code the entire project, and validate results together with the business specialists and decision makers.

12.4 **CASE STUDY**

For this research, a survival/mortality classification of breast cancer women was studied. Clinical cases were used in Mexican-origin women living in the United States. Data source, survival, and mortality rates for breast cancer come from the Surveillance, Epidemiology and End Results (SEER) program of the National Cancer Institute of the United States. SEER is responsible for the national cancer registry and is the main source of information authorized for this disease in the United States. Currently, a variety of research is conducted through the use of these cancer registries, which are available to researchers, physicians, public health officials, politicians, research groups, and the public in general [32].

Information on cancer cases, in particular deaths/survival rates, are of crucial importance when reporting on cancer trends, determining whether prevention and control efforts are effective, identifying possible research directions, and taking action when potential increases are reported of the incidence of cancer.

In a preliminary analysis of data, 740,506 records and 146 variables were identified between 1973 and 2012. In addition, other analyses made from the SEER database were used for this research. These analyses, in Molero et al. [33] and Molero [3], were carried out under statistical models, such as data correlational analysis and principal components analysis, and the opinion of specialists in the field of health, from which 34 significant variables were identified that are directly related to breast cancer and with sufficient records in consecutive periods. From this analysis, vertical (variable) and horizontal (rows) selections of the data were made, with only variables associated with breast cancer taken into account in Mexican-origin women residing in the United States. Thus, the final minable data view was made up of 16 variables and 2652 records (Table 12.2), taking as a class the variable "Vital Status recode," whose values are: zero for mortality (0) and one for survival (1).

For this work, the design solution was implemented using user-centered Web technology (Fig. 12.3). The application consists of four main sections: (a) operators panel, which contains functions to load the data source, select the minable data view, select the data mining algorithm, and validate its accuracy; (b) design panel, which allows the structure of the sequence of the operators for the execution of data mining algorithms; (c) operators configuration panel, which allows operators to be configured in the design section; and (d) results panel, which presents the results obtained through interactive interfaces.

Table 12.2 Variables That Make Up the Minable Data View

Number	Variable Name	Description	Type	Length
1	Marital Status at Dx	Civil status	Discrete	1
2	Age at diagnosis	Age of patient	Discrete	3
3	Month of diagnosis	Diagnostic month	Discrete	2
4	Year of diagnosis	Diagnosis year	Discrete	4
5	Laterality	Side where the tumor originated	Discrete	1
6	Behavior Code ICD-O-3	Type of behavior of the neoplasm	Discrete	1
7	Grade	Classification of cancer cells	Discrete	1
8	Diagnostic Confirmation	Method of confirmation of cancer	Discrete	1
9	Regional Nodes Examined	Removed and examined lymph nodes	Discrete	2
10	Rx Summ-Radiation	Radiotherapy method performed	Discrete	1
11	Rx Summ-Surg/Rad Seq	Sequence of surgery and radiotherapy	Discrete	1
12	Age Recode <1 Year olds	Age group (5-year intervals)	Discrete	2
13	Survival Months	Patient survival time (months)	Discrete	4
14	Tumor Size	Tumor size	Discrete	3
15	AJCC Stage	Stage of disease	Discrete	2
16	Vital Status recode	Patient's state of life	Discrete	1

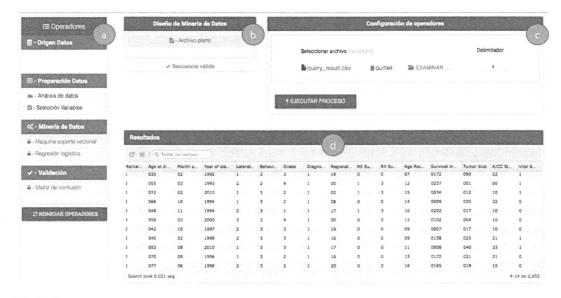

FIG. 12.3

Web application interface of user-centered data mining.

12.4.1 OPERATORS

Operators are the functions that the application uses for management and analysis of the data source, selection of variables, choice of algorithms, validation of results, and presentation of the obtained data patterns. These operators are:

- Flat file. It has the function of loading and displaying the flat type data set, with formats of comma-separated values (csv), text file (txt), and Excel (xls or xlsx).
- Data analysis. This operator serves as the support for analyzing the data set through statistical measurements (average, standard deviation, variance) and graphs in two dimensions of lines, points, and bars.
- Variables selection. Used to configure the role of variables that are part of the minable data view, that is, as input for independent variables and output for one or more dependent variables (class).
- Support vector machine (SVM). Through this operator the predictions of one or more dependent variables are made, taking as input historical data of the independent variables [34].
  The pseudocode of this algorithm is as follows:
  Step 1. Normalize the data $A = x - \mu/\sigma$
  where $\mu = \sum x/n$
  $$\sigma = \sqrt{1/n - 1\sum(xi - \mu)}$$
  Step 2. Compute augmented matrix $[A - e]$
  Step 3. Compute $H = D[A - e]$ and $H^T H$
  Step 4. Compute $U = V[I - H(I/V + H^T H)^{-1} H^T] e$
  Step 5. Compute $w = A^T D U$ and $\gamma = -e^T D U$
  Step 6. Compute $w^T x - \gamma$
  Step 7. Compare the sign $(w^T x - \gamma)$ with input class label
- Logistic regression (LR). This operator is a statistical method, which, like the previous operator, allows the prediction of a class variable based on the predictor variables of the data set [35].
  The function of logistic regression is defined as:
  $$\text{Log}(P/(1 - P)) = B_0 + B_1 X_1 + B_2 X_2 + \ldots + B_n X_n$$
  where P is the probability of occurrence of the event of interest, X represents the independent variables, and B represents the coefficients associated with each variable.
- Confusion matrix. This function evaluates the accuracy of the classification algorithms. The evaluation consists of validating if the predicted value coincides with the actual value (Table 12.3).

where TP are the cases correctly classified and belonging to the positive value, FP are the cases incorrectly classified and belonging to the positive value, TN are the cases correctly classified and belonging to the negative value, and FN are the cases incorrectly classified and belonging to the negative value.

12.4.2 USABILITY AUTOMATON

In order for the user to commit the least number of errors in the use of the application, a usability automaton was designed and implemented to validate the execution sequence of the operators in the design panel. For this, functions were defined to enable and disable operators with the purpose

Table 12.3 Confusion Matrix

		Real	
		Yes	**No**
Classification	Yes	TP (True positive)	FP (False positive)
	No	FN (False negative)	TN (True negative)

of improving user usability and experience. Each of the automaton nodes represents an output, in this case one or more valid or failed sequences. It is important to emphasize that if it were necessary to include new operators in the tool, the automaton allows the appending of new nodes, making the system scalable. The automaton is defined as a tuple (Q,I,d,F) on a single alphabet {A}, where Q is the set of states, I is the initial state, d is a transition relation, and F is the final state. More precisely, states that comprise the automaton are:

- OD1. Access data through text files.
- PD1. Statistical measurements and presentation of data in two-dimensional graphs.
- PD2. Configuration of the role of the variables in the data set (input, output, or none).
- MD1. Support vector machine method.
- MD2. Logistic regression method.
- VD1. Validation of classification methods through a confusion matrix.

The transition relation is defined as follows: $d(OD1,A)=PD1$, $d(OD1,A)=PD2$, $d(PD1,A)=PD2$, $d(PD2,A)=MD1$, $d(PD1,A)=MD2$, $d(MD1,A)=VD1$, and $d(MD2,A)=VD1$. The final state is VD1. Fig. 12.4 depicts a graphical representation of the automaton; Fig. 12.5 describes a graphical representation of an invalid action sequence. An invalid sequence occurs because, once the data set (OD1) is loaded, some data mining methods (MD1 or MD2) are used without having defined, following the type of supervised learning, the role of the variables that make up the minable data view (PD2). By preventing these failed actions, it is possible to reconfigure the variables and the data set to obtain new results.

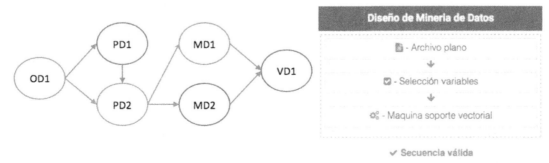

FIG. 12.4

Valid sequence detected by the usability automaton.

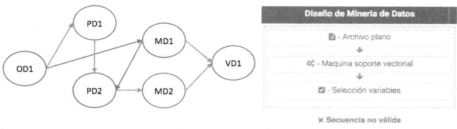

FIG. 12.5

Failed sequence detected by the usability automaton.

12.4.3 USABILITY TESTING

Two-phase usability tests were performed to detect and make improvements to the user-centered data mining tool. In both tests, the method of guided tasks was used: that is, all the tasks that the users had to perform were dictated aloud. As part of the evaluation process, the participants were recorded with video cameras in order to monitor their interaction with the Web application. In addition, they were given a letter of consent and a questionnaire to collect information about their experience when using the tool.

For the initial tests, we worked with eight users with knowledge of data mining. These users were postgraduate students from the technology area of the Universidad Veracruzana. The results of the correct and incorrect tasks performed by the users are shown in Fig. 12.6.

Six users successfully completed all tasks, and another two had an error completing one of the assigned tasks. In addition, videos were analyzed to find usability problems. In the variable selection stage it was found that there was not enough information to perform the task, and corrections were

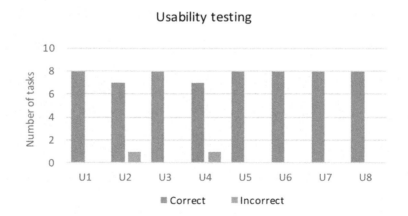

FIG. 12.6

Correct and incorrect tasks performed by users.

made in the tool. Once the application was built and evaluated from the point of view of usability, the execution was performed to analyze the diagnosed cases of breast cancer in Mexican-origin women.

Subsequently, the second phase of evaluation involved four users, oncology specialists. Medical users of the General Hospital La Raza of Mexico City were involved in this phase, based on a cooperation agreement, due to the current work performed by these specialists in the prevention, diagnosis, treatment, control, and epidemiological surveillance of breast cancer. This allowed a greater understanding of the survival rate and mortality of patients with breast cancer and improvements were then made to the data mining tool. The evaluation used the SIRIUS method, proposed by Suárez [36], who defines it as a user-oriented Web usability assessment system based on the determination of critical tasks. From the evaluation it was observed that there is coherence between the usability results achieved by the users (Table 12.4), having high usability values ranging from 87.9% (user 2) to 96.2% (user 3). These values indicate that the appearance and presentation of the Web application are liked by medical specialists, making the application a usable project.

Another method of evaluation used with medical specialists was a usability verification checklist based on the heuristics of Nielsen [37]. Fig. 12.7 shows the values reached in the evaluated criteria:

Table 12.4 Results of the Usability Evaluation by Medical Specialists using the SIRIUS Method	
Users	**SIRIUS Value (Usability Percent)**
User 1	93.9
User 2	87.9
User 3	96.2
User 4	90.9

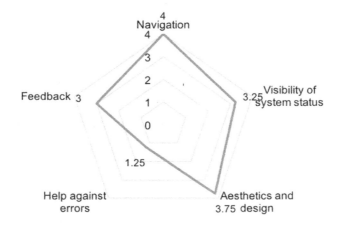

FIG. 12.7

Results of the usability verification based on heuristics.

(a) navigation, (b) visibility of system status, (c) aesthetics and design, (d) help against errors, and (e) feedback.

Based on the results, it was observed that the Web application presents acceptable usability results: that is, in the navigation heuristic, the four users showed a perfect positive satisfaction, with an average of 4. Similarly, a high acceptance, or positive satisfaction, was reached in the heuristics of visibility, aesthetics, and feedback of the application, with values close to 4: that is, 3.25, 3.75, and 3, respectively. The heuristic with less satisfaction was the aid to errors, with a value of 1.25. This result does not represent a critical value preventing the use of the application, but rather an opportunity to include new interactive functions that could serve as a guide or assistance to the nonexpert user, in this case, medical specialists.

12.5 **RESULTS**

After the execution of the classification methods, logistic regression and support vector machine, the results were evaluated through the confusion matrix, whereby classification accuracies of 87.4% and 85.6% were obtained, respectively. This indicates that, with these classification methods, it is possible to achieve truly acceptable forecasts. The accuracy obtained by logistic regression is shown in Table 12.5. It was observed that the predicted values of breast cancer survival and mortality follow a pattern similar to that of the original data. Based on these results, the accuracy of the model and other measures such as the sensitivity and specificity of the nonuniform behavior of breast cancer can be analyzed in a practical manner.

Fig. 12.8 shows similar behavior between actual and predicted cases. An increase of patients with this pathology is also observed, reflecting that breast cancer is a disease with a nonuniform trend, since over time the number of cases detected increased significantly. This may be due to an increase of patients with this pathology, as well as to the extension of surveillance and follow-up of the disease. Consequently, it is well-known that breast cancer is a major and growing public health problem.

On the other hand, Fig. 12.9 shows that the mortality rate gradually declined in the last 3 years of analysis (2010 − 12). This decline may be due to the scientific efforts and advances that have been made in modern medicine to deal with this deadly disease, which continues to lead to the loss of many lives.

Other validity parameters, such as sensitivity and specificity, were calculated. Sensitivity is the probability of correctly classifying a living individual: that is, the probability that a living person is classified as a true positive (survival). In this case, a value of 95% was obtained. Specificity is the probability of correctly classifying a dead person: that is, a death is classified as true negative (mortality), with 72%.

Table 12.5 Results Obtained by the Classification Methods						
Method	Cases	True Positives	True Negatives	False Positives	False Negatives	Accuracy
Logistic regression	2652	1695	624	239	94	87.4%
Support vector machine	2652	1725	547	316	64	85.6%

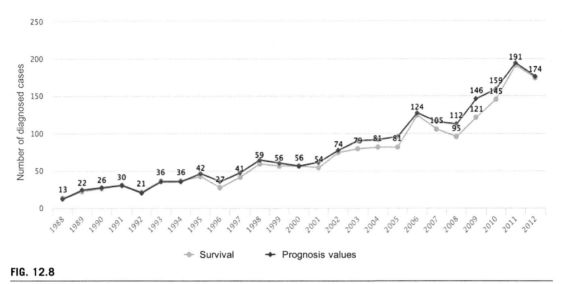

FIG. 12.8

Classification of survival of Mexican-origin women diagnosed with breast cancer.

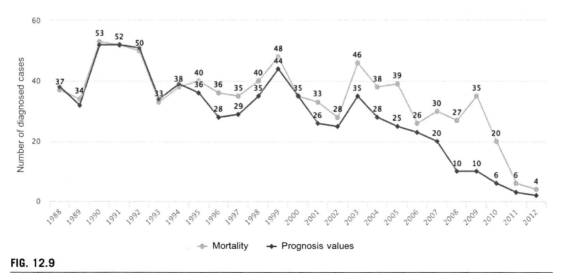

FIG. 12.9

Classification of mortality of Mexican-origin women diagnosed with breast cancer.

$$\text{Sensitivity} = \frac{TP}{TP+FN} = \frac{1695}{1695+94} = \frac{1695}{1789} = 0.947 = 95\%$$

$$\text{Specificity} = \frac{TN}{TN+FP} = \frac{624}{624+239} = \frac{624}{863} = 0.723 = 72\%$$

The higher the sensitivity and specificity value (in a range between 0% and 100%), the better the detection of survival or mortality in patients diagnosed with breast cancer. Based on this, the result has a high sensitivity and an acceptable specificity. This indicates that there is a high probability of predicting cases of survival or mortality of a woman of Mexican-origin diagnosed with breast cancer.

12.5.1 RESULTS PRESENTATION

For the presentation of the results, an interface with interactive graphic functions was implemented, which were defined as functional requirements by the end users [38], medical specialists of the General Hospital La Raza of Mexico City. Fig. 12.10 shows the interaction that the user makes with one of these graphs, related to the side on which the cancerous tumor develops (left, right, both, or unknown), facilitating a better presentation and understanding of the results for the user. These graphics are related to oncological variables of interest defined by medical specialists. These variables are: (a) laterality, side where the tumor originated; (b) behavior code ICD-O-3, behavior of the neoplasia; (c) diagnostic confirmation, method of confirmation of cancer; and (d) RX Summ-Radiation, method of radiotherapy used.

Regarding the behavior of the neoplasm (Behavior Code ICD-O-3), the distribution and variability of the type of breast cancer are shown (Fig. 12.11 and Table 12.6). It was observed that the largest number of cases was concentrated in the malignant type (invasive) with 2357 records; this represents 88.88% of the total cases diagnosed, while the remaining cases, 295, are of the carcinoma in situ (noninvasive) types, representing 11.12%. Of these cases, 65.2% survived a malignant carcinoma, while 85.4% survived carcinoma in situ.

On the other hand, taking into account the best method used for confirmation of cancer (DX_CONF), 97.95% of the cases were confirmed through positive histology (Fig. 12.12 and Table 12.7). In addition, a smaller group of cases were confirmed through positive cytology (12),

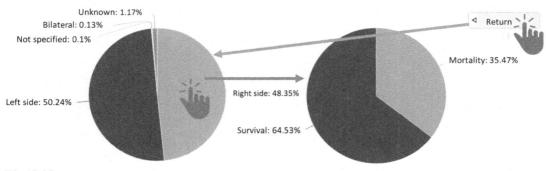

FIG. 12.10

Interactive graphic related to the laterality of the tumor.

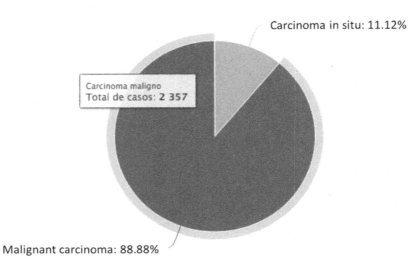

FIG. 12.11

Survival and mortality according to the type of neoplasm.

Table 12.6 Classification of Survival and Mortality According to the Type of Neoplasm				
Neoplasm	**Cases**	**Percentage**	**Survival**	**Mortality**
Malignant carcinoma	2357	88.88%	1537 (65.21%)	820 (34.79%)
Carcinoma in situ	295	11.12%	252 (85.42%)	43 (14.58%)

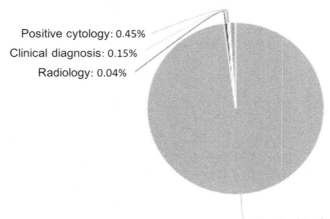

FIG. 12.12

Survival and mortality according to the type of confirmation of the tumor.

Table 12.7 Classification of Survival and Mortality According to the Type of Confirmation of the Tumor

Confirmation	Cases	Percentage	Survival	Mortality
Positive histology	2635	99.36%	1782 (67.63%)	853 (32.37%)
Positive cytology	12	0.45%	7 (58.33%)	5 (41.67%)
Radiology	1	0.04%	0 (0%)	1 (100%)
Clinical diagnosis	4	0.15%	0 (0%)	4 (100%)

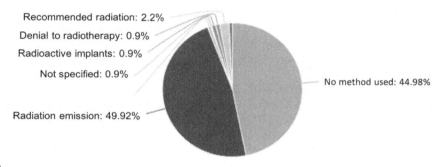

Recommended radiation: 2.2%
Denial to radiotherapy: 0.9%
Radioactive implants: 0.9%
Not specified: 0.9%
No method used: 44.98%
Radiation emission: 49.92%

FIG. 12.13

Survival and mortality according to the type of radiotherapy used.

radiology (1), and clinical diagnosis (4). Of these, it was observed that 67.6% of the cases confirmed by positive histology survived breast cancer.

Another variable of interest was the radiotherapy method used as part of the cancer treatment (RX Summ-Radiation). It was observed (Fig. 12.13 and Table 12.8) that the highest probability of survival was achieved by people who received radiotherapy: of 1324 cases, 74.3% survived. People who did not receive any method, 1193 cases, had a lower probability of survival (59.3%). Another smaller group of cases that refused the radiotherapy emission (24) presented a high percentage of mortality, with 41.7%; this contrasts with those people who received radioactive implants (23), whose survival probability is above 91.3%.

Table 12.8 Classification of Survival and Mortality According to the Type of Radiotherapy Used

Radiotherapy	Cases	Percentage	Survival	Mortality
Radiation emission	1324	49.9%	984 (74.3%)	340 (25.7%)
No method used	1193	44.9%	707 (59.3%)	486 (40.7%)
Recommended radiation	57	2.2%	43 (75.4%)	14 (24.6%)
Denial of radiotherapy	24	0.9%	14 (58.3%)	10 (41.7%)
Not specified	24	0.9%	16 (66.7%)	8 (33.3%)
Radioactive implants	23	0.9%	21 (91.3%)	2 (8.7%)

12.6 CONCLUSIONS

In this chapter, we propose a user-centered classification system for the analysis of health and well-being data. In particular, we described an implementation of a system for the analysis of mortality/survival rates for breast cancer in Mexican-origin women. The system integrates the ISO 9241–210:2010, regarding user-centered features, with the classic data mining process CRISP-DM. The system is composed of three main stages: analysis, data mining, and deployment. The analysis stage is divided into contextual and data analysis, and data preparation. The data produced at this first stage is the input for the data mining stage, which itself is composed of two main phases: modeling and evaluation. The classification resulting from this second stage is finally presented in the deployment stage. User-centered fundamentals are considered at each stage of the system.

Moreover, in this chapter, we also described an implementation of the classification system in a Web application. This application is composed of two classification algorithms (logistic regression and support vector machine), a confusion matrix, and an interactive graphical interface for the presentation of the obtained results.

As a data source, we used the database of the Surveillance, Epidemiology and End Results (SEER) program of the National Cancer Institute of the United States. We first report an initial analysis of the variables. This report is composed of two stages. The first consisted of a preliminary review and analysis of the variables recorded in the database. This was done with the purpose of establishing those relevant according to the period of their records, discarding the variables that presented a high amount of null or missing data. In the second stage, the quality of the data series was determined to establish the significant variables used in the case study classification process. As a result, a minable data view, consisting of 16 variables and 2652 records, was generated.

Regarding the analysis of the survival and mortality of Mexican-origin women diagnosed with breast cancer, it was observed that this cancer is a disease with nonuniform behavior, indicating that the possibility of recovery depends on the cancer stage, whether it is located only in the breast or has spread to other parts of the body, type of treatment received, and the health of the patient in general. Therefore, a diagnosis at an early stage of this pathology is crucial.

For the classification of survival and mortality by breast cancer, the dependent variable was the state of life of the patient, which represents the life or death situation of the woman with breast cancer. Thus, through this variable, it was possible to establish acceptable predictions of the survival and mortality of breast cancer cases based on historical information from a group of independent variables.

From the evaluation of the classification methods, by means of confusion matrices, it was observed that the greater accuracy was reached with logistic regression (87.4%). This indicates that for every 100 cases diagnosed with breast cancer, 87 could be accurately predicted with a history either of survival or mortality. In addition, sensitivity (95%) and specificity (72%) indicate predictions with greater reliability.

The usability tests (SIRIUS and checklist), conducted with users having knowledge of data mining and medical users, served to make improvements in the application. The test results were contrasted with the satisfaction of the end users. This indicated that the operation, appearance, and presentation of the Web application were validated by medical specialists, making it a usable and functional tool. More precisely, the evaluation considered eight users with previous knowledge of data mining tools, and from those, six successfully completed all tasks, whereas only two were not able to complete a single task. Regarding usability tests with experts, the evaluation considered four specialists from the General

Hospital La Raza of Mexico City. For the SIRIUS tests, results ranged from 87.9 to 96.2, out of 100. For the checklist based on Nielsen heuristics, all four users reported perfect satisfaction (4).

There are a several research perspectives for which user-centered data mining can be studied, such as: (a) do more usability testing, involving other users from the medical field; (b) add other algorithms in the Web application for the purpose of developing new projects, not only focused on health, but also on other scenarios; (c) make a comparison with the process defined by Maldonado [39] in order to identify common points and scale this user-centered data mining proposal.

REFERENCES

[1] Hernández J, Ramírez M, Ferri C. Introducción a la Minería de Datos. Madrid, Spain: Pearson Prentice Hall; 2004. p. 680, ISBN: 84-205-4091-9680.

[2] Witten I, Frank E. Data mining: practical machine learning tools and techniques. San Francisco, United States: Morgan Kaufmann Series in Data Management Systems; 2005. p. 525, ISBN: 0-12-088407-0525.

[3] Molero G. Clasificador bayesiano para el pronóstico de la supervivencia y mortalidad de casos de cáncer de mama en mujeres de origen hispano. PhD thesis, Mexico: Universidad de Guadalajara; 2014.

[4] Antonio A, Molero G, Rojano R, Velázquez M. Minería de datos centrada en el usuario para el análisis de la supervivencia y mortalidad de casos de cáncer de mama en mujeres de origen mexicano. Res Comput Sci 2016;123(1):51–63.

[5] Horberry T, Burgess R. Applying a human-centred process to re-design equipment and work environments. Safety 2015;1(1):7–15.

[6] Zhao Y, Chen Y, Yao Y. User-centered interactive data mining. Proceedings of firth IEEE international conference on cognitive informatics, IEEE Computer Society; 2006. p. 457–66.

[7] ISO. Ergonomics of human system interaction—Part 210: Human-centred design for interactive systems. International Standardization Organization; 2010 [ISO 9241-210:2010].

[8] Brachman R, Anand T. The process of knowledge discovery in databases. In: Advances in knowledge discovery and data mining. American Association for Artificial Intelligence; 1996. p. 37–57.

[9] Haun S, Nürnberger A. In: Supporting exploratory search by user-centered interactive data mining. 34th international ACM SIGIR conference on research and development in information retrieval, Beijing, 24–28 July; 2011.

[10] Habib ur Rehman M, Sun C, Ying T. UniMiner: towards a unified framework for data mining. In: Information and communication technologies. IEEE Computer Society; 2014. p. 134–9.

[11] Nigro H, González S, Xodo D. Data mining with ontologies: implementations, findings, and frameworks. Hershey, PA, United States: IGI Global; 2007. p. 3333, ISBN: 978-1-59904-620-43333.

[12] Moine J, Gordillo S, Haedo A. In: Análisis comparativo de metodologías para la gestión de proyectos de Minería de Datos. 17th argentine congress of computer sciences; 2011. p. 931–8.

[13] SAS. Data mining and the case for sampling [online]. Data Mining Using SAS Enterprise Miner; 2015. Available from, http://sceweb.uhcl.edu/boetticher/ML_DataMining/SAS-SEMMA.pdf. Accessed 20 June 2016.

[14] Sumathi S, Sivanandam S. Introduction to data mining and its applications. Studies in computational intelligence, vol. 29. Berlin, Germany: Springer-Verlag; 2006. p. 828, ISBN: 3-540-34350-4828.

[15] KDnuggets. New standard methodology for analytical models [online]. Available from, www.kdnuggets.com/2015/08/new-standard-methodology-analytical-models.html; 2016. Accessed 14 June 2016.

[16] Rivo E, de la Fuente J, Rivo A, García E, Cañizares M, Gil P. Cross-industry standard process for data mining is applicable to the lung cancer surgery domain, improving decision making as well as knowledge and quality management. Clin Transl Oncol 2012;14(1):73–9.

[17] Britos P. Procesos de explotación de información basados en sistemas inteligentes. PhD thesis, Argentina: Universidad Nacional de la Plata; 2008.

[18] Pyle D. Business modeling and data mining. London, England: Morgan Kaufmann Series in Data Management Systems; 2003. p. 650, ISBN: 978-1558606531650.

[19] Jang G, Jeon J. A Six Sigma methodology using data mining: a case study on Six Sigma project for heat efficiency improvement of a hot stove system in a Korean steel manufacturing company. Multiple Crit Decis Mak 2009;72–80.

[20] IMSS. Breast cancer [online]. Mexican Social Security Institute; 2016. Available from, www.imss.gob.mx/salud-en-linea/cancer-mama. Accessed 7 May 2016.

[21] Mukherjee S. The emperor of all maladies: a biography of cancer. New York, United States: Scribner; 2010. p. 592, ISBN: 978-1-4391-0795-9592.

[22] NCI. Cáncer de mama: Información general sobre el cáncer de mama [online]. National Cancer Institute; 2016. Available from, www.cancer.gov/espanol/tipos/seno. Accessed 8 May 2016.

[23] INEGI. Estadísticas a propósito del día mundial de la lucha contra el cáncer de mama [online]. Available from, www.inegi.org.mx/saladeprensa/aproposito/2015/mama0.pdf; 2015. Accessed 7 May 2016.

[24] Secretaria de Salud. Guía de Práctica Clínica: Diagnóstico y Tratamiento del Cáncer de Mama en Segundo y Tercer nivel de Atención. México: Centro Nacional de Excelencia Tecnológica en Salud; 2009. p. 102.

[25] ACS. Causes, risk factors, and prevention topics [online]. American Cancer Society; 2016. Available from, www.cancer.org/cancer/breastcancer/detailedguide/breast-cancer-risk-factors. Accessed 10 June 2016.

[26] ASCO. Cáncer de mama: Síntomas y signos [online]. American Society of Clinical Oncology; 2015. Available from, www.cancer.net/es/tipos-de-c%C3%A1ncer/c%C3%A1ncer-de-mama/s%C3%ADntomas-y-signos. Accessed 6 May 2016.

[27] ASCO. Cáncer de mama: Factores de riesgo [online]. American Society of Clinical Oncology; 2015. Available from, www.cancer.net/es/tipos-de-c%C3%A1ncer/c%C3%A1ncer-de-mama/factores-de-riesgo. Accessed 10 May 2016.

[28] NCI. Cáncer de seno (mama)—Versión para profesionales de Salud [online]. National Cancer Institute; 2016. Available from, www.cancer.gov/espanol/tipos/seno/pro. Accessed 8 June 2016.

[29] Ho T, Nguyen T, Nguyen D, Kawasaki S. Visualization support for user-centered model selection in knowledge discovery and data mining. Int J Artif Intell Tools 2001;10(4):691–713.

[30] Moré A. MPIu+a Ágil: El modelo de proceso centrado en el usuario como metodología ágil. MSc thesis, Universidad de Lleida, Spain; 2010.

[31] Abras C, Maloney D, Preece J. User-centered design. In: Bainbridge W, editor. Encyclopedia of human-computer interaction. vol. 37(4). Thousand Oaks: Sage Publications; 2004. p. 445–56.

[32] NCI. Surveillance, epidemiology, and end results program [online]. National Cancer Institute; 2016. Available from, https://seer.cancer.gov/data/. Accessed 10 July 2016.

[33] Molero G, Céspedes Y, Meda M. Caracterización y análisis de la base de datos de cáncer de mama SEER-DB. 9th international congress on health informatics; 2013. p. 1–9.

[34] Cortes C, Vapnik V. Support-vector networks. Mach Learn 1995;20(3):273–97.

[35] Park D, Kim H, Choi I, Kim J. A literature review and classification of recommended systems research. Expert Syst Appl 2012;39(11):10059–72.

[36] Suárez M. SIRIUS: Sistema de evaluación de la usabilidad Web orientado al usuario y basado en la determinación de tareas críticas. PhD thesis, Spain: Universidad de Oviedo; 2011.

[37] Nielsen J. Usability engineering. San Francisco, United States: Morgan Kaufmann; 1993. p. 361, ISBN: 978-0-12-518406-9361.

[38] Antonio A. Proceso de minería de datos centrado en el usuario con base a la norma ISO 9241-210:2010. MSc thesis, Mexico: Universidad Veracruzana; 2016.

[39] Maldonado G. Minería de datos centrada en el usuario bajo un enfoque en el proceso de ingeniería de la usabilidad y accesibilidad. MSc thesis, Mexico: Universidad Veracruzana; 2016.

MODELING INDEPENDENCE AND SECURITY IN ALZHEIMER'S PATIENTS USING FUZZY LOGIC

13

Meza-Higuera Jesus A. *, Zamudio-Rodriguez V. Manuel*, Doctor Faiyaz[†], Baltazar-Flores Rosario*,
Lino-Ramirez Carlos*, Rojas-Dominguez Alfonso*

National Technology Mexico/Leon Institute of Technology, Leon, Mexico* University of Essex, Colchester, United Kingdom[†]

13.1 INTRODUCTION

According to Alzheimer Mexico, Alzheimer's disease occupies fourth place in progressive diseases that do not produce immediate death. This disease slowly causes complete dependence on assistance from others in patients, and ultimately those suffering from Alzheimer's become unable to perform their own activities. Alzheimer's disease worsens over time; in its early stages, memory loss is mild, but in advanced stages, people lose the ability to maintain a conversation or even respond adequately to their environment.

Adults who present with this dementia will need care commensurate with the extent of disease progression. Alzheimer's disease will severely affect the person over time, with memory loss, difficulty solving problems and performing habitual tasks at home or work, disorientation to time and/or place, problems in the use of words or writing, decreased judgment, and mood swings. Regardless of the Alzheimer's stage that a person presents in, that person will require constant physical and mental health monitoring.

In an intelligent environment, one of the important requirements is the need to know the user's location within the smart environment. In recent years progress has been made in indoor localization systems, but most of them still pose problems, such as spending excess time in the calibration process and nonrobustness and high cost of installing new hardware equipment [1]. We can use available technologies to help take care of adults in intelligent environments, such as GPS, camcorders, microphones, smartphones, etc. These mainly allow monitoring or tracing of the person within the intelligent environment, for purposes ranging from analysis of the person's behavior in the intelligent environment, to safety and care implementation.

The primary caregiver is the person who habitually attends and satisfies the needs of the patient: this can be the spouse, a child, a niece, a nurse (paid caregiver), or another family member [2]. The caregiver should always look for effective ways to care for the person, either directly or indirectly.

Caregivers will have the extra task of caring for their own physical and mental integrity while taking care of a patient. The caregiver must have strong physical and mental health. They may suffer from exhaustion, stress, insomnia, or changes in their appetite or behavior. Caregivers can put their own health at risk by not adequately caring for themselves as well. The proposed system can help to alleviate caregiver burden.

13.2 FUZZY LOGIC

Fuzzy logic is the theory of fuzzy sets, or sets that calibrate vagueness. Fuzzy logic is based on the idea that all things admit degrees (Figs. 13.1 and 13.2). Take as examples temperature, height, speed, distance, beauty: these all fall along a mobile scale. Such a mobile scale often makes it impossible to distinguish members of a class from nonmembers. (For example, when does a hill become a mountain?) Boolean or conventional logic uses sharp distinctions, and forces us to draw lines between members of a class and nonmembers.

(A) (B)

FIG. 13.1

Range of logical values in fuzzy and Boolean logic. (A) Boolean and (B) Diffuse [3].

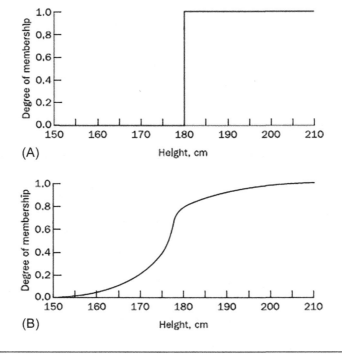

(A)

(B)

FIG. 13.2

Graphical sets of Boolean logic (A) and fuzzy logic (B) [3].

Formally, fuzzy logic is determined to be a set of mathematical principles for the knowledge of representation based on degrees of membership and not on the clear membership of classical binary logic.

13.3 FUZZY MODEL

The project described in this chapter focuses on an aid for adults with Alzheimer's dementia. The chapter discusses why this disease was studied first, the behavior of the people who suffer from it, the help that exists today, and the state of the art related to the objective of this project.

The system presented is based on fuzzy logic models that have been investigated in the technical literature. Their general functioning has been studied, including designing fuzzy sets, membership functions, processing of data, and fuzzy rules. The system described aims to allow decision making (rules) concerning the safety of a user with Alzheimer's, considering their independence and the level of Alzheimer's present. It also gives the caregiver an additional tool to help with patient safety monitoring. Using fuzzy logic helps to provide a balance between security and independence; for this, we have defined the variables of entry and exit with their respective functions of membership.

The variables and membership functions were modeled using the Juzzy tool [4], designed by Christian Wagner. The input–output variables, membership functions, and the graphic of the given functions were obtained from this tool.

The Juzzy tool internally performs the processes of diffuse inference and defuzzification [5]. Also, the calculations of the outputs of the modeled system were handled by Juzzy, but on the desktop version the output values are calculated by the Java programming tool. Thus, the system will pass the input data and process it using Juzzy, and the output value will be sent to and used in the simulator.

As mentioned, any model based on fuzzy logic should provide a balance between security and independence. For this purpose, the variables of input and output were defined, with their respective functions of membership.

The system architecture discussed here is not developed in a single model; rather, it is regarded from a point at which it is the union of small models, which work together to offer more flexibility. It was judged that a single model would be a more complex and complicated way to meet the objectives. Instead, by adding simpler models, all the requirements are expected to be covered.

Each particular model will be called at the time it is needed, and then will send its results to another model. This work between models is handled through various combinations, either by detecting abnormal scenarios in the environment or by performing a calibration of the parameters of the system.

The model consists of four input and four output variables, each of which has a different functional purpose in the fuzzy system. Each has different values; these values are elements that are not normally measurable, or their value depends on different opinions, such as "much," "little," "very far," or "too close." Also, each variable has its own working universe: in other words, the range of numerical values that each accepts (see Tables 13.1 and 13.2).

13.4 CALIBRATION

The system incorporates several types of adjustments called calibrations. These adjustments are made so that the models can recalibrate the output variables. In other words, certain settings will decide when to change the system's security or independence levels.

Table 13.1 Input Variables

Name	Description	Values	Universe
Alzheimer's stage	The stage of dementia that the person with Alzheimer's is in	Early stage Mid stage Advanced stage	(1 ... 7)
Alzheimer's score	Indicates the score obtained in an aid tool oriented to the cognitive state of the person with Alzheimer's	Very high High Regular Low Very low	(1 ... 100)
Time	The inactive time taken on the patient	No time Short time Regular time Long time	(1 ... 100)
Repetition	Reflects the occasions in which the person repeats an activity or movement	No repeat Little repeat Regular repeat High repeat	(1 ... 10)

Table 13.2 Output Variables

Name	Description	Values	Universe
Security	Indicates the level of security that the person requires	Very weak Weak Regular Restricted Very restricted	(1 ... 100)
Independence	Indicates the level of freedom to be given to the person in the environment	Very free Free Regular Restricted Very restricted	(1 ... 100)
Wandering	Indicates when the person is conducting a random and aimless route	Low frequency Frequent Very frequent	(1 ... 10)
Downtime	Indicates when the person enters a state of inactivity	Little Moderate Alarming	(1 ... 10)

Calibration of the fuzzy models allows the system to make decisions regarding the environment. Thanks to detected events, the system is thus able to change its current parameters and adapt to the needs of the person.

Five calibrations have been developed for system operation. The initial calibration serves as the system's starting point. The different calibrations will not always occur in the same order, as the order will depend on the person being monitored. Calibrations are performed when certain patterns are detected, or when certain data are collected from the environment.

Each calibration also has fuzzy rules for operation within the simulator. These rules are those that interpret the values to be given in each calibration, according to the linguistic crosses between variables and their inputs or outputs.

13.4.1 INITIAL CALIBRATION

This calibration will adjust security and independence levels according to the values of the variables of the stage of Alzheimer's and Alzheimer's score. This calibration is the initial calibration of the system.

13.4.2 INDEPENDENCE CALIBRATION

This calibration will adjust the independence level based on the variables of stage Alzheimer and repeat. This calibration must be carried out with the increment of the repeat variable, according to the place where the adjustment was made. When the person visits some places more frequently than others, that place will be marked as a recurring site. The place will be marked as restricted or allowed, depending on the Alzheimer's stage that the person is in.

13.4.3 TIME-INACTIVE CALIBRATION

This calibration is initiated when the person is inactive, based on the time of inactivity. If inactivity is detected, the system must recalibrate the safety of the person and alert the caregiver. The time will be measured only when the person is in a place and is not moving; this is not the same as the time that passes in a place, like the time variable.

13.4.4 AMBULATION CALIBRATION

This is initiated when the model detects the person entering a state of wandering, thanks to the variable of time and the repetition of the person entering the same zones within the environment.

13.4.5 SAFETY CALIBRATION

This adjusts the safety level according to the variables of stage Alzheimer and downtime. This calibration must be carried out when the system detects that the person needs more restricted security, based on the time passed as motionless.

13.5 EXPERIMENTATION AND RESULTS

Once agreement on simulation of the fuzzy model was reached, we looked for different tools that could be helpful. We found the UbikSim tool developed by Emilio Serrano et al. [6], which is an intelligent and ubiquitous environment simulator. UbikSim makes use of multiagent-based simulation in intelligent environments (AI).

The scenarios were created using the UbikEditor tool, which comes within the source code of the UbikSim project. The stage developed has one floor with multiple rooms, such as a garden, bathrooms, and living room, and also doors and sensors of presence and movement.

The first scenario performed is the initial calibration: when you start the simulation, you must choose and mark the primary and secondary places on the stage. In other words, these places are where the system should pay more attention in its monitoring. The initial calibration will be the starting point of the model, and the model should reflect the initial data on the 3D draft of the environment. This will be done according to the type of primary/secondary zones, risk (represented by green, yellow, and red colors), the name of the zone, number of zones, and the initial calibration. The risk level, presented in colors, has the following meanings: green—zone without risk; yellow—zone with moderate risk; and red—zone of high risk. In Fig. 13.3 we can see an example of a stage with initial calibration and four zones (two primary and two secondary). To show the use of the initial calibration, some tests will be carried out with multiple start-up parameters. The procedure is as follows: the primary or secondary zones are selected manually by the stage name; these are given the value of 0 or 1 respectively, for main or secondary zone; and finally the initial calibration is captured using the Alzheimer's stage and Alzheimer's score.

There are four tests, consisting of testing the initial calibration with multiple input values. Each test has different calibration parameters shown in a table. These calibration values were input manually on the simulator code. The parameters are: zone numbers, which are the number of types of zones that the test scenario will have. Zones are the room's names and are used on the test scenario for monitoring. Alzheimer Stage and Alzheimer Score are the values for the initial calibration and starting point of the simulator.

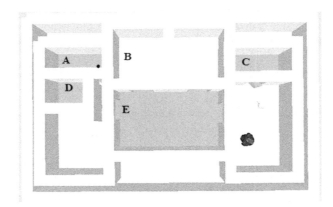

FIG. 13.3

First test initial calibration.

Table 13.3 Test 1 Initial Calibration Parameters

Zone Numbers	2 Primary/1 Secondary
Zones	Kitchen (D), Living Room (E), Bathroom 1 (A), Bathroom 2 (C)
Alzheimer stage	1
Alzheimer score	9

The first test uses data from a person with Alzheimer's disease in its early stages. The parameters are shown in Table 13.3, and the result is reflected in Fig. 13.3.

Fig. 13.3 shows the A, C, D zones on zones without risk ; this is because the initial calibration data reflects a person with the early stages of Alzheimer's.

The second test uses data from a person with Alzheimer's disease at an intermediate stage. The parameters are shown in Table 13.4, and their result is reflected in Fig. 13.4.

Fig. 13.4 shows the A, C, D zones on zones with moderate risk; this is because the data for the initial calibration reflects a person suffering from the intermediate stages of Alzheimer's disease.

The third test uses data from a person with Alzheimer's disease in the latest stages. The parameters are shown in Table 13.5, and their result is reflected in Fig. 13.5.

Fig. 13.5 shows the A, C, D zones on zones with risk; this is because the data for the initial calibration reflects a person suffering from the last stages of Alzheimer's.

Table 13.4 Test 2 Initial Calibration Parameters

Zone Numbers	2 Primary/1 Secondary
Zones	Kitchen (D), Living Room (E), Bathroom 1 (A), Bathroom 2 (C)
Alzheimer stage	4
Alzheimer score	5

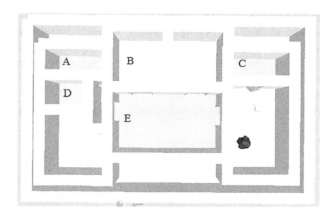

FIG. 13.4

Second test initial calibration.

Table 13.5 Test 3 Initial Calibration Parameters

Zone Numbers	2 Primary/1 Secondary
Zones	Kitchen (D), Living Room (E), Bathroom 1 (A), Bathroom 2 (C)
Alzheimer stage	7
Alzheimer score	1

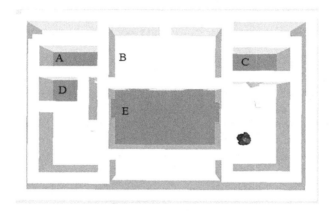

FIG. 13.5

Third test initial calibration.

The fourth test uses random data from a person with Alzheimer's disease. The parameters are shown in Table 13.6, and their result is reflected in Fig. 13.6.

A more accomplished experiment is the detection of abnormal situations on the stage, during which the calibrations and adjustments are activated given certain situations, such as detection of nonmovement of the person, ambulation, or frequency of activities. With the simulated scenario running, the system will detect and notify when some particular calibration is required and applied. In the following examples of calibration detection, hard code was used, i.e., these events were forced, according to each case. Since the scenario is normally tested with random behavior detection, it would be very difficult to perceive the exact behavior required for a sample.

The following examples show when the system detects an anomaly and performs a calibration in addition to notifying it.

Table 13.6 Test 4 Initial Calibration Parameters

Zone Numbers	2 Primary/1 Secondary
Zones	Kitchen, Living Room, Bathroom 1, Bathroom 2
Alzheimer stage	6
Alzheimer score	3

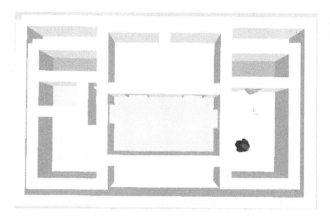

FIG. 13.6

Fourth test initial calibration.

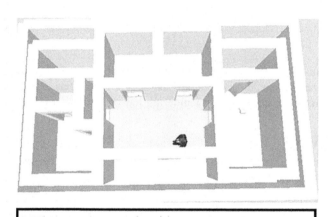

INFO: Abnormal time is detected in the place: living room.

INFO: Applying safety calibration

FIG. 13.7

Safety calibration.

13.5.1 **SAFETY CALIBRATION**

This calibration is activated by the Alzheimer's stage and the time that passes within a certain place without activity (Fig. 13.7).

Depending on the Alzheimer's stage, a threshold time is set, after which the system will detect the inactivity as abnormal and security will be modified.

13.5.2 **WANDERING CALIBRATION**

This calibration is driven by repetition of entering the same places within a short period of time (Fig. 13.8).

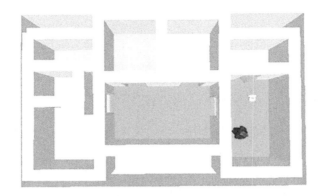

INFO: Patient has entered the state of wandering.

INFO: Notifying caregiver.

FIG. 13.8

Wandering calibration.

This calibration is initiated by the person's route within the area. If the person follows the same route frequently in the same areas, the system detects that the person has entered a state of wandering. Detecting this will notify the caregiver that wandering has occurred.

13.5.3 TABLE OF RUNS

A series of experiments were carried out with the system in which each experiment began with different values. The method used was as follows:

- Starting values were given to the initial calibration, where E is equal to Alzheimer's stage and P is equal to Alzheimer's score. Values 1, 3, 5 are the numbers of the value of each variable. For example, the value $E = 1$ is the first value of the variable Alzheimer stage for Early_Stage.
- The detection fields note the calibrations that the system detects and in what order. Three detection fields were recorded, as the system has detected a maximum of three calibrations in a single run.
- Detection values are:
  - The name of the calibration is entered in bold.
  - The rule that was activated at that time is below the type of detection (each rule was investigated in the fuzzy system code at the time of Detection).
- The fuzzy system reached an end of execution, which is reflected in the output field. The output values are usually:
  - Restricted area, when the person in the environment enters an area marked as restricted and the system ends the execution to warn the caregiver.
  - Wandering, when the person enters a state of wandering; usually the system will stop the execution when detecting wandering on constant occasions, or when the stage of Alzheimer is high.

- Expected output notes when the system should not have given that output, as for example in the second run, the system detected wandering but didn't perform a calibration; instead the execution was finished, so the system did not act as it should (Table 13.7).

Different results were obtained in the previous table, but this emphasizes that the system detects wandering more frequently, since its calibration appeared in almost all the executions. Errors were presented at the time of ambulation detection: the system should have passed to the wandering calibration, but instead simply detected and finished the simulation.

The system has two ways to end an execution when it is not done manually: one way is by detecting the person in a restricted zone (that zone is determined by running the program, using hard code adjustments), or the other way is when an alarming degree of user wandering is detected.

Table 13.7 Executions of the Fuzzy System Model

	Initial Values	Detection	Detection	Detection	Output	Expected Output
Values	$E=1$ $P=1$	– 	–	–	Restricted Area	–
Values	$E=1$ $P=3$	-Wandering – –	–	–	Wandering	Wandering Calibration
Values	$E=1$ $P=5$	-Wandering Short_Time Short_Repetition	-Security Short_Time Early_Stage	–	Restricted Area	–
Values	$E=2$ $P=1$	-Security Alarming_Time Mid_Stage	–	–	Restricted Area	–
Values	$E=2$ $P=3$	-Independency Mid_Stage Lot_Repetition	-Wandering Short_Time Few_Repetition	-Wandering Regular_Time Regular_frequently	Wandering	–
Values	$E=2$ $P=5$	-Security Short_Time Mid_Stage	–	–	Restricted Area	–
Values	$E=3$ $P=1$	-Independency Advance_Stage Short_Repetition	-Wandering – –	–	Wandering	Wandering Calibration
Values	$E=3$ $P=3$	-Security Short_Time Advance_Stahe	-Security – –	–	Restricted Area	Security Calibration
Values	$E=3$ $P=5$	-Wandering No_frequently Short_Time	-Independency Advance_Stage No_frequently	–	Without Output	Detection Calibration

In the Output field, there were correct outputs, which ended up in the restricted area or wandering, but there were also some exit errors. However, in some executions the system simply opted to end the simulation and did not continue with the corresponding action that in some cases should have been the ambulation calibration and then continue with the simulation.

In the last execution, the system simply stopped the simulation; when checking the source code at the time of the output, nothing out of the ordinary was noticed, but the correct output should have been to continue with the simulation and call a calibration. This error took place only in one execution.

When the systeam were initialize with low values, it should call calibrations when high fuzzy rules are detected, like in the case of the run example that have initial values of $E = 2$ and $P = 1$, where the security calibration was called with the value Alarming_Time. This is because the system must detect danger in low initial values when the data tooked from the patient are elevated. However, if the initial values are high, the system should call calibrations at the time of detecting low data values from the patient, such as with the initial values $E = 3$ and $P = 5$, where the security setting was called with the value of Short_Time, because high initial values means that the patien are on the lastest stages of the dementia and every detection or move should be watched.

13.6 CONCLUSIONS

The system presented in this chapter aims to have a person with Alzheimer's under constant monitoring of activities engaged in (at this stage, in a simulated environment). By means of fuzzy logic, the system also responds according to the information collected. Tests were carried out using the Ubiksim platform, which allows simulation of complex social behaviors in both 2D and 3D environments. A scenario was implemented with behavior inspired by a test case documented by the Alzheimer's Foundation of America. This first experiment showed that the system follows the patient's movements within the environment, as it detects both cases of the tests developed. The first occurred when the person moved randomly in the environment, reflecting wandering. Detecting wandering in the system was not a simple problem; just because the person enters different rooms does not mean that he/she is wandering. We approached the detection of the problem from another perspective: the wandering would be detected when the system detect that the patien its entering the same room over and over and also perfoming the same action.

The simulator works and responds according to the fuzzy logic model. The membership functions of the model must adjust the monitoring environment, and the second experiment was based on the response given by the system with the first calibration or membership function. At the beginning, the system must capture the initial data, which in this case are the values of the secondary areas, and the variables of Alzheimer_Stage and Alzheimer_Score. When capturing the data and before the test scenario is generated, the system makes the corresponding calculations in the initial calibration and gives the areas determined as main and secondary their colors, which varies depending on the result of the calibration.

Notification of the calibrations appears when the corresponding behaviors are detected. In the environment of testing with random behavior, it is not likely to obtain the calibration desired, so in the experiments the calibrations were called by manipulating the source code, to show that the system makes use of calibrations when required.

The model allows detection of wandering by detecting events in the simulator. An event is a programmed pattern for which the simulator will register when the situation occurs. For example, when the person spends time without moving, the system will detect that pattern, register it, and perform the corresponding actions, as well monitoring various other parameters, including position. In addition, the developed model allows detection of situations of risk for the patient, such as being in an area near the entrance/exit door of the house. These tests show the efficiency of the model, presented in a preliminary way, in preventing the patient from becoming lost, especially during the night. Using different activities, extensive experiments are currently being carried out, considering patients with different levels of Alzheimer's (low/high, with random data), different areas of risk (low, medium, high), as well as scenarios for both day and night.

The logic-based model provides a balance between independence and security for people with Alzheimer's in a friendly environment. This helps both people with Alzheimer's and their caregivers to feel calmer within the environment, as the model increases patient monitoring, notifying the caregiver when a risk situation occurs. The model works through the relevant variables in the environment, using fuzzy inference rules for the safety of the user and considering the patient's Alzheimer's level as well as the design of test scenarios that allow the model to be validated.

The system follows the track of the person, detects which rooms are entered, and the time that is spent in them. The environment has both main and secondary rooms; zones change in degree or color when the system makes adjustments to the environment. The system helps the caregiver through notification when the person is in risky situations, such as entering dangerous areas, entering a state of wandering, random night travel, among others.

The model is able to give extra help to the caregiver, decreasing some of the workload and stress in assisting the patient with Alzheimer's. Help will be through an alert notification when the model detects a risk situation in the person with Alzheimer's. As the patient moves through the different attitudes and degrees of dementia, the system can help maintain balance between appropriate independence and security. The modeling of the system is developed through fuzzy logic. This model was conceived from the perspective of the security and independence of a person with Alzheimer's. We chose the necessary factors for this system as the time that the person is motionless, in a state of inactivity; when the person starts to wander in the environment without fixed course; and the degree of repetition in movements in the environment (such as the number of times in a row that the person goes into the same room over and over).

The model has its limitations, such as lack of physical sensors or other modules that could provide more patient information. However, it is possible to add to or improve the model. Also, the detection in the simulator is not quite correct, as there are times when the system does not behave properly. More modules can be added in future to cope with different situations. Also, as future work, the model is compatible with the addition of new fuzzy models for better performance. Another possible implementation in the future is to employ the system in a physical environment, with sensors and appropriate technologies. The model is ultimately intended to work with real Alzheimer's patients and real behavior and movements in an intelligent environment.

REFERENCES

[1] Bradford D. Detecting degeneration: monitoring cognitive health in independent elders. Proc 35th Annual International Conference of the IEEE Engineering in Medicine and Biology Society (EMBC), Osaka 2013;7029–32. https://doi.org/10.1109/EMBC.2013.6611176.

[2] Cook DJ. Health monitoring and assistance to support aging in place. Univers Comput Sci 2006;12(1):15–29.

[3] Navarro J, Zamudio V, Doctor F, Lino C, Baltazar R, Martinez C, et al. Game based monitoring and cognitive therapy for elderly. Proceedings of the 9th international conference on intelligent environment 17: 2013. p. 116–27.

[4] Lin JM, Lin CH. RFID-based Wireless Health Monitoring System Design, Proc Eng 2013;67:117–27. https://doi.org/10.1016/j.proeng.2013.12.011.

[5] Van Leekwijck W, Kerre EE. Defuzzification: criteria and classification. Fuzzy Sets Syst 1999;108 (2):159–78.

[6] Zhou Z, Dai W, Eggert J, Giger JT, Keller J, Rantz M, He Z. A real-time system for in-home activity monitoring of elders. Proc 2009 Annual International Conference of the IEEE Engineering in Medicine and Biology Society, Minneapolis, MN 2009; pp. 6115–8. https://doi.org/10.1109/IEMBS.2009.5334915.

FURTHER READING

[1] A. F. of America. Lost and found—a review of available methods and technologies to aid law enforcement in locating missing adults with dementia [online], Available: https://www.alzfdn.org/documents/Lost&Found_forweb.pdf; 2012.

[2] Wagner BC. Juzzy—a Java based toolkit for type-1, interval type-2 and general type-2 fuzzy logic and fuzzy logic systems; 2013. p. 13–4.

[3] Cook DJ, Youngblood M, Heierman III EO, Gopalratnam K, Rao S, Litvin A, et al. In: MavHome: an agent-based smart home. Proceedings of the first IEEE international conference on pervasive computing and communications, 2003 (PerCom 2003); 2003. p. 521.

[4] Serrano E, Botia J. Validating ambient intelligence based ubiquitous computing systems by means of artificial societies. Inf Sci 2013.

[5] Medjahed H, Istrate D, Boudy J, Baldinger JL, Bougueroua L. A fuzzy logic approach for remote healthcare monitoring by learning and recognizing human activities of daily living; 2015.

[6] Tung J, Snyder H, Hoey J, Mihailidis A. Everyday patient-care technologies for Alzheimer's disease; 2013. p. 80–3.

[7] Zadeh LA. Information and control. Fuzzy Sets 1965;8(3):338–53.

[8] Negnevitsky M. Artificial intelligence: a guide to intelligent systems. 1st ed. Boston, MA, USA: Addison-Wesley Longman Publishing Co., Inc.; 2001.

[9] Paper P. Technologies for remote patient monitoring for older adults. Center for Technology and Aging; 2010. April.

[10] Oliveira R, Barreto A, Cardoso A, Duarte C, Sousa F. Environment-aware system for Alzheimer's patients. Proc. 4th Int. Conf. Wirel. Mob. Commun. Healthc.—Transforming Healthc. through Innov. Mob. Wirel. Technol; 2014. p. 8–11.

[11] Paiva S, Abreu C. Low cost GPS tracking for the elderly and Alzheimer patients. Procedia Technol 2012;5:793–802.

[12] Mandal SN, Choudhury JP, Chaudhuri SRB. Search of suitable fuzzy membership function in prediction of time series data. Int J Comput Sci Issues 2012;9(3):293–302.

[13] Garcia-valverde T, Garcia-sola A, Hagras H, Member S, Dooley JA, Callaghan V, et al. A fuzzy logic-based system for indoor localization using WiFi in ambient intelligent environments. IEEE Trans Fuzzy Syst 2013;21(4):702–18.

[14] Chen YM, Cheng KS. GPS-based outdoor activity pattern recording and analysis system. Proc. Annu. Int. Conf. IEEE Eng. Med. Biol. Soc. EMBS, no. 1; 2013. p. 1160–3.

WIRELESS SENSOR NETWORKS APPLICATIONS FOR MONITORING ENVIRONMENTAL VARIABLES USING EVOLUTIONARY ALGORITHMS

14

Lino-Ramirez Carlos, Zamudio-Rodriguez V. Manuel, Ochoa-López Verónica del Rocio, Muñoz-López Gerardo

National Technology Mexico/Leon Institute of Technology, Leon, Mexico

14.1 INTRODUCTION

The IEEE 802.15.4 standard is an interconnection protocol for personal area wireless networks with low data transmission rates, LR-WPAN (low rate-wireless personal area network). This type of network uses low-speed data devices, low power consumption, and short-range radio frequency transmissions. They are characterized by the connections through the WPAN, where there is little or no structure, unlike WLAN-type networks. In wireless sensor networks (WSNs), signals are transmitted using radio frequency. The IEEE 802.15.4 standard defines four possible frequency bands in which the sensor nodes can operate. The choice of frequency band depends on factors such as interference, restriction of spectrum use, and coexistence with other systems, among others. However, in the transmission system, it is also necessary to take into consideration the efficiency of the antenna, because the nodes only use small antennas, which means a decrease in the wavelength. If the efficiency—that is, the power of effective transmission—decreases, more energy will be needed to reach the desired level [1].

The access layer to the media provides two services: MAC Data Service, which allows the transmission and reception of data from the MPDU (MAC Protocol Data Units), and MAC Management Service, which is unfolded as the interface of the layer management entity MAC (MLME) with the Access Service (SAP) or also MLME-SAP [1]. The MAC protocol has the task of managing access to the communication medium. Among the main functions of the MAC layer are: the management of the beacon (which is a small device that emits a wireless broadcast signal), channel access, guaranteed time slot management or GTS, validation of macros, association, and dissociation [1].

There are currently several MAC protocols. However, the most widely used are CSMA (Carrier Sense Multiple Access) and MACA (Multiple Access with Collision Avoidance), in addition to its

Intelligent Data Sensing and Processing for Health and Well-being Applications. https://doi.org/10.1016/B978-0-12-812130-6.00014-7

variations. CSMA is an algorithm based on containment-based protocols for wireless sensor networks. In this algorithm, in its nonpersistent form, a wireless node is authorized to immediately transmit the data once the media is released; if the media is busy, a rollback version occurs and starts a timer to retry the process. On the other hand, in persistent CSMA, a node that wants to transmit data continuously scans the media to find possible activity. When the media is free, the node transmits its data; if there is any collision, the node will wait a random time before making another try to transmit [2]. CSMA/CA (CSMA Collision Avoidance) is a variation of CSMA. This MAC protocol improves collision evasion. When the media is free to transmit, the nodes do not transmit immediately. Rather, they generate a waiting period and compare their times. Thus, the node with the smallest waiting period will be able to transmit [2].

14.1.1 SENSOR NODE ARCHITECTURE

Wireless sensor nodes are the central elements of a WSN. Each node is responsible for obtaining information, processing it, and communicating or transmitting it to its neighbors, and it also stores and executes communication protocols and data-processing algorithms. The capacity and performance of the network will depend largely on the attributes of the nodes, so it is of utmost importance to select their characteristics according to the needs of the application. Following, the structure of the nodes and the elements that compose them are described.

14.1.1.1 Hardware components

A basic sensor node comprises five main components, as shown in Fig. 14.1: controller, memory, power devices, communication devices, and a power supply [3].

14.1.1.1.1 Controller

The controller is a driver to process all relevant data, capable of executing arbitrary code [3]. A microcontroller is a small integrated circuit, usually composed of a central processing unit, memory, parallel input and output interfaces, a clock generator, one or more analog-to-digital converters, and serial communication interfaces. They are recommended for stand-alone applications, since they offer programming flexibility [2]. There are other devices that can be used as controllers, such as digital signal processors (DSPs), application-specific integrated circuits (ASICs), programmable logic devices

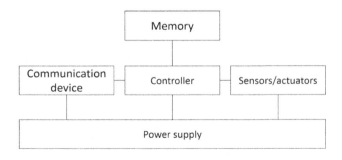

FIG. 14.1

Components of a sensor node.

(PLDs), or field programmable gate arrays (FPGAs). DSP processors perform discrete signal processing through digital filters, which reduce noise or modify the signal spectrum to optimize it. These processors are highly capable of processing millions of samples per second in real time. They are recommended for multimedia applications, or applications in hostile environments where the signal can be affected by noise and interference [2]. ASIC processors are designed according to the final application, and FPGA processors adapt to needs. These require energy and time, so both types of processors are costly [3].

14.1.1.1.2 Memory
Some sensors need memory to store programs and data. In general, different types of memory are used: RAM, ROM, or EEPROM. Due to power restrictions in wireless sensor networks, it is important to consider the time required in read and write operations, due to the amount of energy needed to perform them [3].

14.1.1.1.3 Sensors and actuators
Different types of sensors can be used in a WSN. These are classified into three categories:

- *Passive, omnidirectional sensors*: These sensors can measure a physical magnitude; they have only one active prove.
- *Passive, narrow-beam sensors*: These sensors are passive and have a well-defined idea of the direction of the measurement. A typical example is a camera, which can "take action" in a given direction and allows you to rotate if necessary.
- *Active sensors*: This last group of sensors actively probes the environment; for example, a sonar or radar sensor or some types of seismic sensors, which generate shock waves using small explosions.

As with sensors, a variety of actuators can be used, but for a WSN one must bear in mind that, in general, the sensor node can only open/close a switch [3].

14.1.1.1.4 Communication
The communication device is used to exchange data between individual nodes. In some cases, cable communication is the method of choice and is often applied in many network-like environments. Wireless communication can be carried out by means of radio frequencies, optical communication, or ultrasound. Other media such as magnetic inductance are only used in very specific cases [3]. For true communication, both a transmitter and a receiver are required on a sensor node. The essential task is to convert a bitstream from a microcontroller to radio waves. For practical purposes, it is generally advisable to use a device that combines these two tasks into a single entity. Such combined devices are called transceivers.

To select appropriate transceivers, a number of features must be taken into account. The most important are

- Service to the top layer: The transceivers must provide an interface that allows the MAC layer to initiate the raster transmissions and packet delivery. In the other direction, incoming packets must be heard in buffers accessible via the MAC protocol.
- Energy consumption and energy efficiency: This refers to the energy necessary to transmit and receive a single bit. In addition, the transceivers must be able to differentiate between the different states, for example, active or sleep states.

- Carrier and multiple frequency channels: Transceivers must be available for different carrier frequencies, in order to alleviate network congestion problems [3].

14.1.1.1.5 Power supply

Power to feed the sensor nodes can be supplied from batteries, solar cells, etc. [3]. The means of feeding each sensor is directly proportional to the life of the network, so be careful when selecting a protocol that optimizes its use to the maximum. It is important to consider the storage and supply of energy that is required, in addition to the replenishment of energy consumed [2].

14.1.1.2 Software components

Sensors usually have five basic software subsystems, as follows (see Fig. 14.2) [4]:

- *Microcode or middleware operating system (OS)*: This is the most common software on the WSN boards, used by all the node-resident software modules to support various applications. The purpose of an operating system is to isolate or protect the software from the machine-level functionality of the microprocessor. For this reason it is necessary to have open operating systems specifically designed for WSNs. These OSs generally use an architecture that allows for quick implementation with a minimum code size.
- *Sensor controllers*: These are software modules that manage basic functions of sensor transceivers; these sensors can be of the modular/plug-in type, depending on the type and sophistication and the appropriate settings. The settings must be loaded to the sensor or sensors.
- *Communication processors*: These software subsystems manage the functions of communications, including routing, packet buffering and sending, maintenance topology, media access control, encryption, and FEC, to name a few.

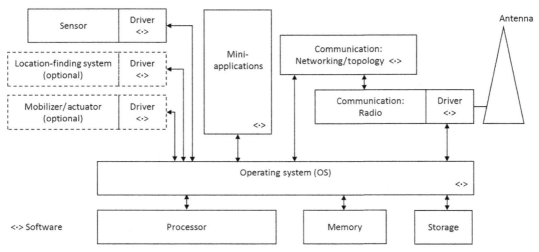

FIG. 14.2

Software components of a WSN.

- *Communication controllers*: These modules manage the details of the transmission channel of the radio link, including timing and synchronization, signal coding, bit recovery, bit count, signal levels, and modulation.
- *Small data-processing applications*: These are numeric applications, data processors, signal-level storage and data manipulation, or other basic applications supported at node level for network processing.

14.2 DESIGN AND IMPLEMENTATION OF A WSN ARCHITECTURE FOR A WATERING APPLICATION

This section proposes a network architecture for automated greenhouse irrigation, for greater water savings. For this purpose, a centralized architecture was designed, which consists of the following elements: sensor nodes in charge of gathering information about temperature and humidity variables in the environment and the soil; actuator nodes that allow the flow and supply of water to the plants; an intelligent node that analyzes the information obtained and runs an optimization algorithm that establishes irrigation and wait times between each irrigation; a server that stores the data collected by the system; and a visual interface that accesses that data to show it as graphs to the user. It also has a central node in charge of communication between nodes and actuators. IEEE 802.15.4 technology is used as the standard underlying the Zigbee protocol for wireless communication, as it is capable of supporting mesh routing that allows data packets to cross multiple nodes (multiple "hops"), with the purpose of investigating the target node and making it possible for Zigbee nodes to extend over large regions and continue to support communication between all devices on the network [5].

The elements that make up the proposed architecture are described here in more detail:

Central node: This node is responsible for providing dynamic network addresses to its child devices, that is, to the sensor and actuator nodes that make up the network. The central node continually receives information from each of the nodes and in turn communicates with the smart node in order to send the information to it, so the central node cannot enter low-power mode. The central node consists of an Xbee module, which connects to the smart node using an Xbee adapter through a serial interface protocol.

Sensor nodes and actuators: These are nodes formed by sensors that measure ambient temperature and humidity, as well as soil moisture. These sensors are connected to an Arduino One board, which reads the values of the sensors and transmits the information through an Xbee antenna to the central node. The node also has a rechargeable battery that allows it to operate in places where access to an outlet is not possible. All these components are contained in a plastic box that protects them from the dust and water inside the greenhouse.

Sensors: The sensors used are devices that measure humidity and ambient temperature, in addition to the humidity of the soil, with characteristics as described in Table 14.1.

Table 14.1 Sensor Nodes Characteristics

Sensors	Microcontroller	CI	Operating Range	Value
Temperature	Arduino ONE	DH11	−40°C to 125°C±0.5°C	6
Ambient humidity			0% to 100% with 5% accuracy	
Soil moisture		SEN92355P	0% to 100%	

The sensors are configured to send information to the coordinator about the percentage of soil moisture, humidity, and ambient temperature, in addition to sending the number of the group of plants to which they belong. The coordinator is in charge of receiving the information and, depending on the origin of the information, to check if the humidity levels are in the range set for the group and, if not, to send a signal to the node that controls the actuators responsible for providing water to the group that requires water. This allows water to be provided to achieve an acceptable level of humidity for that group. Soil moisture sensors are placed in the roots of the plants, because the most important thing is to maintain the moisture of the roots at an acceptable level and thus allow favorable plant growth. Once the sensor gets the information and sends it to the network coordinator, the data is stored in a database within the server, from which they are read with a periodicity given by an application developed in Java, responsible for displaying the behavior of variables since the system started until the last reading received by the nodes.

Actuators: Actuators are solenoid valves that allow irrigation to be supplied to plants within the greenhouse. For the construction of each actuator, a solenoid valve connected to an Arduino One plate was used, which in turn is connected to an Xbee module. The solenoid valve works with a 12-V power supply. The actuators operate as follows: the solenoid valve is adapted to a water inlet provided with a hose containing a set of pipettes, by means of which the water flows individually towards each of the plants to moisten the soil. Actuators are activated when the central node sends a signal individually to each one indicating that the plants to which the actuator supplies water need to be irrigated. The actuators are programmed with the activation times provided by the optimization algorithm; in this way only the necessary quantity is supplied in the periods of time that allow it to maintain the humidity of the plants at an acceptable level.

Configuring sensor, actuator, and central node nodes: Communication between the sensor nodes and actuators with the central node is carried out using the Zigbee communication protocol. Each of the Xbee modules that each node has is configured in API (Application Programming Interface) mode. To configure the modules, X-CTU software was used. Fig. 14.3 shows the setting.

The sensor and actuator nodes are configured as routers, so they act as child devices, because they need a parent (central node) to provide them with their dynamic network address and associate them with a network. However, they also act as a parent device and can provide network addresses to other child devices (other routers):

Smart node: Processes the information received by the sensor nodes and, on the basis of the results, decides when to activate the actuator nodes. If so, sends a message to the central node, which is in charge of communicating with these nodes to allow the flow of water.

Server: This module has the function of storing the data obtained and sent by the intelligent node. The database was created using the database manager My SQL.

Monitoring interface: The system has a graphical interface for monitoring the behavior of the variable sensing, which consists of the following elements:

- Visualization of sensor nodes: Displays the sensor nodes that make up the network, where the user can view and select those from which variable sensing information is desired (temperature, ambient humidity, and soil humidity).
- Graphics panel: This panel consists of three graphs that show the behavior of the variable sensing independently; each graph shows the value of the variable and the number of readings sent over a period of time by each of the nodes.

FIG. 14.3

Configuring nodes in X-CTU.

- Status of actuator nodes: The interface shows the actuators that are responsible for controlling the opening and closing of the solenoid valves, which are responsible for supplying water to the plants that require it. When the valves are turned off, the box of the group to which they belong is shown in red; it is switched to green until the necessary water has been supplied to the plants of its group.

14.2.1 IMPLEMENTATION OF THE APPLICATION IN A CASE STUDY

The proposed system aims to make efficient use of water for irrigation in greenhouses. It was essential to be able to verify the functionality of this system by implementing it in a real environment, which is described in this section.

14.2.1.1 Scenario description

The proposed system was implemented in a real greenhouse, used for the production of plants of different species with the objective of putting them on sale to obtain income that helps with the expenses generated by an institution supporting the elderly. With the implementation of this system, it is intended to reduce the costs generated by the consumption of water in this activity. The greenhouse has more than 50 different species of plants, which were originally distributed by size and irrigated in the same way (two times a day). The plants are arranged in rows in front of a water inlet equipped with a hose; however, these outlets were in disuse and the irrigation was carried out using showers.

14.2.1.2 Use of pipettes for irrigation

First, the use of pipettes was proposed to carry out irrigation. These allow uniform irrigation for all plants in the greenhouse, thus ensuring a water supply for each of them, unlike the use of showers or hoses, where moisture after watering is not uniform for every plant.

14.2.1.3 Plant clustering

The model proposed to group the different species of plants existing in the greenhouse in clusters by common characteristics of soil moisture for each group. This allows a smaller number of groups and better plant retention. The same row of pipettes will be used for all those plants belonging to the same group, and thus the group will be irrigated in the same way, taking advantage of the space and resources of the greenhouse, such as the water intakes used for irrigation. The first step in forming the groups was to gather daily information on the behavior of the soil moisture of the different plant species. A sample of each species was irrigated to a maximum level and a daily sampling of moisture behavior was taken for four or more days, always taking care not to bring the plants to a water stress level.

   The data collected allowed us to learn the behavior of the humidity of the plants, which could be plotted, similarly to the data shown in Fig. 14.4.

14.2.1.4 Setting data and generation of features vector

Once the information on soil moisture in the different plant species was collected, it was proposed to make a data adjustment with the information on each plant to obtain a function describing the curve formed by the behavior of the humidity in each of them. This allows the plants whose curves are similar in shape and position to be in the same group. It is proposed to carry out different regressions and select the function that best fits the data (see Fig. 14.5). The selection criterion of the function to be used to

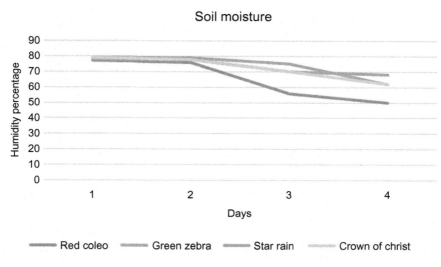

FIG. 14.4

Behavior of moisture in different plant species.

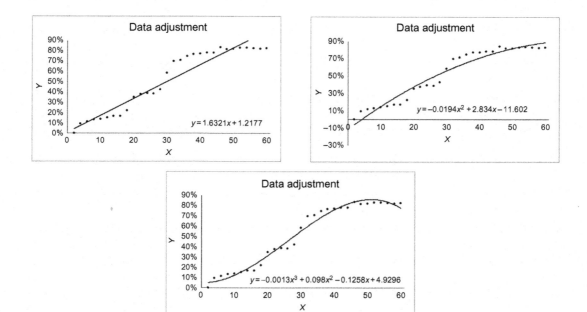

FIG. 14.5

Set-up using different functions.

Table 14.2 Features Vector of Each Type of Plant Formed by the Coefficients of the Equation

Type of Plant	Function Obtained With the Adjustment	Characteristic Vector
Cebra verde	$-1.1227x^2 - 0.1177x + 81.9114$	$\{-1.1227, -0.1177, 81.9114\}$
Coleo Tinto	$-4.515x^2 + 8.185x + 72.83$	$\{-4.515, 8.185, 72.83\}$
Corona de Cristo	$-1.1647x^2 + 7.2326x + 69.9462$	$\{-1.1647, 7.2326, 69.9462\}$
Enredadera	$-0.7388x^2 - 0.2602x + 78.0343$	$\{-0.7388, -0.2602, 78.0343\}$
Estrella (suculenta)	$-0.1737x^2 - 2.2758x + 80.5614$	$\{-0.1737, -2.2758, 80.5614\}$
Estrellita (suculenta)	$-1.2466x^2 + 7.2908x + 71.6076$	$\{-1.2466, 7.2908, 71.6076\}$
Helecho	$-1.5096x^2 + 4.1695x + 77.338$	$\{-1.5096, 4.1695, 77.338\}$
Lavanda	$-6.15x^2 + 13.73x + 67.02$	$\{-6.15, 13.73, 67.02\}$
Lluvia de Estrella	$-1.7493x^2 + 2.3653x + 79.534$	$\{-1.7493, 2.3653, 79.534\}$
Pata de Elefante	$-0.9683x^2 + 4.7946x + 75.0186$	$\{-0.9683, 4.7946, 75.0186\}$
Real de Oro	$-2.1066 + 8.02x + 70.967$	$\{-2.1066, 8.02, 70.967\}$
Rosita (suculenta)	$-0.6408x^2 - 1.3901x + 79.5857$	$\{-0.6408, -1.3901, 79.5857\}$
Santolin	$-1.0099x^2 + 3.6741x + 76.2862$	$\{-1.0099, 3.6741, 76.2862\}$

obtain the values of the coefficients as vectors of characteristics is the one with the smallest average quadratic error.

The values of the resulting function coefficients are considered as the elements of the characteristic vector. With the following function:

$$f(x) = a_1x_1 + a_2x_2 + a_3x_3...a_nx_n \tag{14.1}$$

the features vector will be given by

$$\text{vector} = \{a_1, a_2, a_3...a_n\}$$

The values of $a_1, a_2, a_3...a_n$ are considered to be the features vector of the plant.

The degree of the function must be the same for all plants to ensure a features vector of the same size. Table 14.2 shows the equation obtained with each type of plant, as well as the features vector used in the clustering process.

This method of extracting characteristics is simple and sufficiently effective for this study, since the vectors obtained allowed the formation of groups of plants with similar irrigation needs.

14.2.1.5 Cluster creations and resetting plants

Once the characteristic vector for each plant was generated, the K-means grouping method was used, which groups the different species into clusters whose soil moisture behavior is similar. K-means was used because it is an algorithm of unsupervised learning that allows the number of groups desired to be formed. This is of great importance since the greenhouses have a limited number of water intakes on which the plants could be placed, so the number of groups could not exceed the number of spots available. With the implementation of the clustering algorithm, the goal was that all the curves (which represent the behavior of the humidity) similar to each other would be grouped in the same cluster. After obtaining the groups, the species of plants that have been grouped in the same cluster could then be placed in the same row of pipettes, which will be connected to a solenoid valve allowing the flow of water at the same time to each of the plants belonging to that group.

14.2.1.6 Characterization of formed groups

The characterization consists of determining a group of equations that describes the behavior of the following variables: soil moisture, soil volume, pot size, and sensor position, with each in relation to the watering time (solenoid valve activation time) and waiting time to reach the minimum humidity. Such behavior is different for each group. The system is able to determine the amount of water needed for the plants regardless of the size, the volume of soil they contain, and the position of the sensor in the pot, as the characterization obtained previously allows the optimization algorithm to use the generated data and, based on it, to determine the time of irrigation and wait time. To accomplish this, it is necessary that the user enter these values before implementing the optimization algorithm. To obtain the functions that describe the behavior of the aforementioned variables, a data adjustment is carried out using different mathematical functions with different degrees, and the one whose average quadratic error is least is selected.

The resulting functions at this stage are later used as models of the system behavior in the metaheuristic optimization algorithm, to obtain optimal valve activation and waiting times for each group. These functions describe the behavior of the following relationships:

- Watering time function vs humidity achieved: Describes the behavior of humidity with respect to the time in seconds that the solenoid valve was kept active.
- Watering time vs volume function: Describes the time in seconds necessary to keep the solenoid valve active to achieve the desired humidity in pots with different soil volumes.
- Function watering time vs height: Describes the time in seconds necessary to keep the solenoid valve active to achieve a desired humidity in pots with different sizes.
- Function watering time vs sensor position: Describes the time in seconds that the solenoid valve was activated to achieve the desired humidity in different positions of the pot.
- Function timeout vs humidity reached: Describes the time in seconds after the watering until reaching the minimum humidity set for the group.
- Standby time vs volume function: Describes the time in seconds after watering to reach the minimum humidity in pots with different soil volumes.
- Standby time vs height function: Describes the time in seconds after watering until reaching the minimum humidity in pots of different sizes.
- Function timeout vs sensor position: Describes the time in seconds elapsed after watering to reach the minimum humidity in different positions of the pot.

In order to obtain a function that describes the behavior of each of the aforementioned relationships, plants are used in pots with different volumes of soil, different sizes, and with the sensor in various positions. With each of these variants the moisture behavior is obtained, which is represented by a mathematical function. The system uses the values of volume, size, or height and position of the sensor to establish the necessary irrigation time according to each of the previously defined relationships that will cause a percentage of moisture in the soil in a pot with such characteristics. In the same way it will set the required waiting time for the humidity to reach a minimum humidity. The times obtained will allow an average humidity that could reach the plant from the inputs of volume, height, and position of the sensor to be determined. Fig. 14.6 shows the characterization process to obtain the irrigation time (difference TR, watering time) that allows an acceptable humidity, considering the variables mentioned, to be obtained.

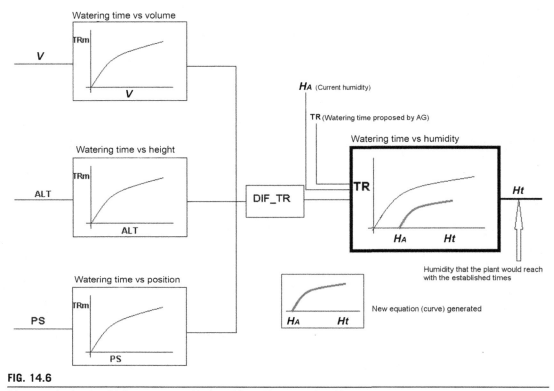

FIG. 14.6

Characterization to obtain an acceptable humidity with a watering time.

The irrigation and waiting times established in the characterization stage, as well as the irrigation and waiting times defined by the optimization algorithm, will allow new equations to be generated from our base equations, which describe the behavior of the humidity with respect to the watering time and the humidity with respect to the waiting time. With the irrigation times of the characterization and those generated by the optimization algorithm, equations are generated that represent new curves placed under our base curve, so the optimization algorithm seeks to obtain the curve that allows an acceptable humidity for the plant group to be maintained. Similarly, the waiting times of the characterization and those generated by the optimization algorithm will generate equations that represent new curves placed above our base curve, so the optimization algorithm seeks to obtain that curve that allows an acceptable humidity for the group of plants to be maintained.

14.2.2 DESCRIPTION OF THE PROPOSED MODEL

To achieve minimum water consumption, it should be considered that the minimum irrigation time is sought as well as the maximum waiting time between each irrigation, in addition to looking for the maximum possible humidity allowed in the plants.

Next, the designed model is described:

- The metaheuristic algorithm begins to optimize times.
- Get the current humidity level of a group of plants using a soil moisture sensor.
- If the humidity difference of the current soil with the humidity calculated by the algorithm is found to be within the accepted window, then the algorithm will stop.
- The optimization (through the metaheuristic algorithm) comes from the characterization of the plants of each defined group. With the characterization, an average humidity can be obtained from a time of irrigation (TR) and the humidity of the soil sensing (HA), an average moisture that could have the plant (Ht). On the other hand, for the determination of the volume consumed of water, given the irrigation times (TR and TE), the following equations can be established:

$$V = T^*Q \tag{14.2}$$

$$f = \frac{T_{PE}}{T_E} \tag{14.3}$$

$$V = f^*(Vtr) \tag{14.4}$$

where Vtr is the volume consumed in watering time, TR is the watering time, Q is the flow of solenoid valve, T_{PE} is the preset time (1 week), T_E is the waiting time between each watering, f is the irrigation frequency, and V is the volume used for a period of time T_{PE}.

Evolutionary algorithms were used for optimization, in this case a genetic algorithm, memetic algorithm and immune system. It is intended that each individual be formed by a string of bits, where half of the chain represents the waiting time and the other half the time of irrigation. In this method, in which the metaheuristic algorithm generates the individuals by a pseudorandom engine, these individuals are tested by the objective function represented by the following equation:

$$F = abs(H_A - Ht)^*V \tag{14.5}$$

So, what is intended is to minimize the objective function.

The solution obtained by the algorithm is then used to adjust the times that the solenoid valve remains active, allowing the water flow and the desired humidity for the plants to be reached, as well as the time that it must remain deactivated until the next irrigation.

14.3 AIR QUALITY MONITORING BY A WSN

The traditional air quality monitoring system is extremely expensive, time-consuming, and energy consuming, and can rarely be used for real-time air quality reporting. Wireless sensor networks are a new alternative to solving this problem in a highly challenging field of research for the automation of embedded systems design, as these designs must meet strict restrictions in terms of power and cost. The proposed system makes use of an air quality index (AQI), which is currently in use. The nodes measure environmental pollution, and they were deployed uniformly [6]. They also generated a scenario simulation with X and Y position of the nodes, through the use of the network simulator (NS-2.33). For better energy management, low-power strategies and a hierarchical routing protocol were used, which caused inactive nodes to stay in sleep status [6]. The Air Pollution Wireless Monitoring System consists of two parts:

- Data: The reading collected by the source node.
- ID, which identifies the node uniquely on the network, such as a network address.

The cluster head or group collects readings from each node and stores them in a list. Three processes are performed to produce a performance output before the WSN model can analyze the information [6]:

- Creation of scenario model
- Simulation
- Analysis

All these processes are performed by the network simulator. When it is running, you can see how the packets flow through the network. The results of the simulation are the number of packets of data sent and received by each node. Performance and delay can be observed [6].

The location of each sensor node n is determined and defined in terms of coordinates (Xn, Yn). The distance (D) between each sensor node of the receiving node was determined by Mishra et al. [6]:

$$D = \sqrt{(Xn)^2 - (Yn)^2} \tag{14.6}$$

The following strategy was used to implement the WSN for this system:

- First, the region of interest is partitioned into several smaller areas to improve the management of the large amount of data collected from the system and better coordination of the different components involved.
- A cluster or group head in each area is implemented, which will form clusters with the nodes in their respective areas, collect data from them, perform aggregation, and send these to the base station.
- The sensor nodes are scattered randomly in the different areas. These get the data and send them to the header or head of the group.
- Multiple base stations are used to collect data from cluster heads or clusters and transmit to entry door (Gateway).
- The gateway collects the results of the base stations and transmits them to the central database [6].

This system uses an air quality index to categorize the different levels of air pollution. It also associates significant and very intuitive colors of the different categories. Therefore the state of air pollution can be communicated to the user very easily. The system also uses AQI to evaluate the level of health concern for a specific area. The wireless air pollution monitoring system uses a novel technique for data aggregation to address the challenge of minimizing energy consumption in the WSN. It also uses quartiles to summarize a list of readings of any length to three values, which reduces the amount of data to be transmitted to the base station, thus reducing the required transmission power and, at the same time, representing the value originals accurately. The system uses a vector-distance algorithm to calculate the maximum permissible distance between sensor nodes with minimal power consumption, which helps deploy the sensor node in a given area. The line graph allows the user to see the tendency of air pollution during several areas at once and the location of the sensors and readings collected by them. This project offers a way to collect data in real time accurately and efficiently [6].

14.3.1 **WSN MONITORING AIR QUALITY WITH LIBELIUM**

Today, the largest percentage of air pollution comes directly from road traffic and no longer from large industries. Road traffic is considered to be responsible for 25% of all emissions in Europe, increased to 31% in Spain only, and 90% of all transport emissions are due to road traffic. The loss of environmental quality is one of the greatest threats of our century, involving human health and well-being, in addition to the environmental impacts. To monitor the quality of the air a network of wireless sensors, Waspmote, was implemented in Salamanca, Spain [7]. The World Health Organization (WHO) says that 12% of the European urban population lives in areas with urban bottom (nontrafficking) concentrations of NO_2 higher than WHO levels. In the city of Salamanca, a rescue project was implemented, financed by the European Union through the LIFE program. The ubiquitous sensor network is currently being tested in the city of Salamanca, where the expected positive impact is the reduction of air pollution levels, to improve health and the human environment [7]. The proposal consists of 35 Waspmotes, which were deployed in two different places, with seven measurement parameters:

- Temperature
- Relative humidity
- Carbon monoxide (CO)
- Nitrogen dioxide (NO_2)
- Ozone (O_3)
- Noise
- Particle 2.5

These seven sensors are connected to Waspmote through a special sensor panel made for this project, which contains the electronic components needed to implement easy hardware integration on these sensors. This sensor plate has been specifically designed to meet the requirements of this project, as Libelium offers custom hardware design. If any of these seven parameters goes above a threshold, then the system analyzes the information and can react by sending an alarm to the central node (MeshLium in this case). To find out where this sensor is, each Waspmote can integrate a GPS, which provides the exact position and time. For accurate measurement, sensors have been calibrated in professional labs, to provide the best results for this project. As one of the partners (CARTIF) says: "The biggest problem they face in this project is related to the calibration of sensors, as there are no companies available to carry out this process." To solve this inconvenience, they "have developed their own calibration methods." These nodes are powered by a solar panel and an external controller that is connected to a battery. Taking advantage of the energy-saving characteristics of Waspmote and solar energy, these specks are autonomous. To assist in data transmission, two Meshliums were also installed, collecting information and sending it through GPRS [7].

14.3.2 **WSN SIMULATION FOR MEASURING AIR QUALITY**

The simulation was carried out in Matlab with the objective of evaluating the behavior of the system under established and controlled conditions. Four experiments were carried out. The simulation parameters for each experiment are provided in Table 14.3. Because the X and Y coordinates are randomly chosen for each node, the four experiments are repeated 10 times to minimize the influence of random deployment. In the end, the averaged results of LEACH and SEP were used for the comparison.

Table 14.3 Simulation Parameters

Parameter	Experiment 1	Experiment 2	Experiment 3	Experiment 4
Size of the área	1000*1000	1000*1000	1000*1000	1000*1000
Position	(500 m, 500 m)	(0 m, 500 m)	(500 m, 500 m)	(0 m, 500 m)
Number of nodes	100	100	1000	1000
Initial energy of nodes (J)	0.5	0.5	0.25	0.25
E_{elec} (nJ/bit)	50	51	52	53
E_{fs} (pJ/bit/m^2)	10	10	10	10
E_{mp} (pJ/bit/m^4)	0.0013	0.0013	0.0013	0.0013
E_{DA} (nJ/bit/signal)	5	5	5	5

For the development of this work, it was first necessary to carry out an analysis of the LEACH (Low Energy Adaptive Clustering Hierarchy) and SEP (Stable Election Protocol) protocols, which allow the transmission of information from one node to the sink, very important protocols for the transmission of the data in the WSN.

LEACH is a protocol that proposes the use of node clusters, allowing the distribution of energy consumption more evenly among the sensors in the network. This protocol is characterized by randomly rotating the cluster header role. Within a cluster, the headend coordinates data transmission and fusion locally to compress the information and then transmit it directly to the base station, thus reducing the bandwidth required.

SEP was an improvement over LEACH in the way that it took into account the heterogeneity of networks. In SEP, some of the high energy nodes are referred to as advanced nodes and the probability of advanced nodes to become cluster header is as compared to that of advanced nodes. SEP does not require any global knowledge of energy at every election round. The drawback SEP method is that the election of the cluster heads between the two types of nodes is not dynamic, which results that the nodes that are far away from the powerful nodes will die first.

14.3.2.1 Simulation environment

In the simulated WSN, 100 homogeneous sensors are randomly deployed in a field of (100 m × 100 m). In an experiment with 100 nodes, the parameters of the first column of Table 14.3 were used, where the communication is through the cluster headers with the base station. In Fig. 14.7, the times of the network life between LEACH and SEP are compared.

The number of rounds before the first node dies in the LEACH protocol and the SEP Protocol is 2. Nodes begin to die faster in the LEACH protocol than the SEP protocol. The number of rounds before the last node dies on the network using the LEACH protocol is less than the number of rounds before the last node dies on the network using the SEP protocol. The number of rounds completed before the last node dies in LEACH and SEP is 1596 and 2001, respectively.

The life of the network, defined as the elapsed time until the last node dies, increased by 25%. The percentage differences in the network lifetimes are calculated using the following equation:

$$\text{Percentage difference} = \left| \frac{r\text{SEP} - r\text{LEACH}}{r\text{LEACH}} \right| *100 \tag{14.7}$$

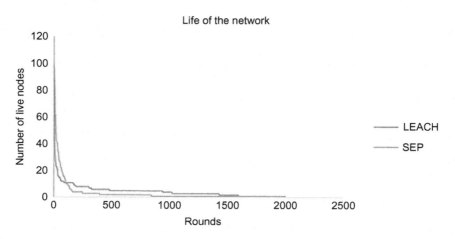

FIG. 14.7

Network life.

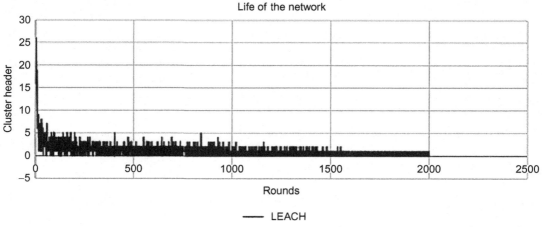

FIG. 14.8

LEACH network life.

Figs. 14.8 and 14.9 show the number of cluster heads selected in each round for the LEACH and SEP protocols, respectively. The optimal number of cluster heads is set to 20 clusters for the two algorithms. Fig. 14.8 shows that in some LEACH rounds up to 25 cluster heads are selected, which is far from the desired number. However, as shown in Fig. 14.9, in the SEP protocol the maximum number of cluster heads to be selected in a round is 25, which is closer to the desired number of cluster heads, and in LEACH is 26. In other rounds, the cluster number selected in LEACH converges faster to one, because most nodes have died.

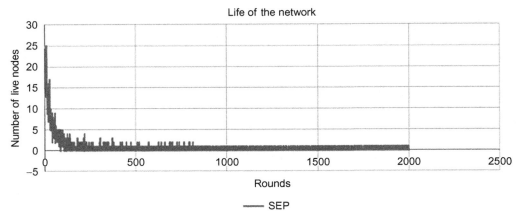

FIG. 14.9

SEP network life.

14.3.3 LIBELIUM WNS IMPLEMENTATION

The sensor used to monitor the air pollution of León City was the Libelium Smart Environment Pro (Fig. 14.10).

Before using the sensors to obtain the data, it was necessary to make a small adjustment to the sensors. The characteristic equation of each sensor was obtained. This process is carried out as follows: Using the graph of the sensor used, Figaro TGS2442, it is possible to see that this sensor presents a logarithmic variation of its resistance according to the concentration in parts per million (PPM) of carbon monoxide (and very light of hydrogen), as can be seen in Fig. 14.11 [8].

According to this graph, we have the following equation:

$$\log(R) = a^* \log(CO) + b \tag{14.8}$$

where:

- R is the sensor resistance (R_s/R_o in the graph),
- CO is the concentration in PPM,
- While a and b are coefficients obtained from the calibration.

R_o has been calibrated at the following way: with a value of 30 PPM in normal conditions, where $R_s/R_o = 4$ and $R_s = 10$ kohms, and $V_{cc} = 5$ V and $V_s = 10$ V, the measurement circuit is calculated with the following equation [8]:

$$R_s = [V_{cc}^* R_L / V_s] - R_L \tag{14.9}$$

Replacing the value of $R_s/R_o = 4$; we obtain $R_o = 40.6$ kohms.

The coefficients a and b of the equation are calculated with the concentration and the resistance corresponding to that concentration. Substituting these values in the equation, we are left with:

$$\log(R) = -1.151 \log(CO) + 2.302 \tag{14.10}$$

In Table 14.4 the characteristic equations of the sensors are observed.

FIG. 14.10

Smart Environment Pro.

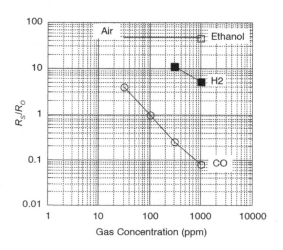

FIG. 14.11

CO sensor response.

Table 14.4 Table of Equations Characteristic of the Sensors

Pollutant	Equation
CO	$\log(R) = -1.151 \log(CO) + 2.302$
NO_2	$\log(R) = -1.737 \log(NO_2) + 2.176$
Temperature	$100 \; X \; -0.5$
PM_{10}	Does not need
SO_2	Does not need
O_3	$\text{Log(ppb)} = -1.1428 * (R_o/R_s) + 1.3859$

Table 14.5 Sensing Data Table

Year	Month	Day	Hour	O_3	SO_2	NO_2	CO	PM_{10}	Temp
2015	December	1	0	17.24	10.53	35.68	1.07	76.92	20.98
2015	December	1	1	21.74	9.52	33.08	1.01	61.7	20.52
2015	December	1	2	4.18	12.11	42.9	1.7	67.08	20.24
2015	December	1	3	13.82	10.58	30.55	0.94	118.73	19.14
2015	December	1	4	17.06	11.31	26.82	0.83	60.05	17.98
2015	December	1	5	19.99	9.99	24.73	0.75	38.16	17.3
2015	December	1	6	18.45	9.57	25.51	0.78	33.1	16.89
2015	December	1	7	11.39	9.83	29.25	0.94	47.34	16.6
2015	December	1	8	3.82	10.58	33.68	1.46	79.21	16.95
2015	December	1	9	6.72	10.83	36.1	1.69	122.28	19.22
2015	December	1	10	25.84	11.65	33.66	1.2	120.73	22.13
2015	December	1	11	41.77	10.77	38.78	1.16	78.02	23.51
2015	December	1	12	64.4	10.8	38.09	0.99	100.26	24.71
2015	December	1	13	58.71	10.38	23.76	0.74	92.1	26.51
2015	December	1	14	58.62	10.09	20.29	0.79	72.09	28.04

With this configuration, sensing began for 1 month and 744 datapoints were obtained, as shown in the example in Table 14.5.

14.4 RESULTS

14.4.1 STATISTICAL ANALYSIS

At this stage, a statistical study was carried out to determine the correlation between variables. Eq. 14.11 was used. Each day there were 24 datapoints. For this reason this process was carried out per day, resulting in Tables 14.6 and 14.7.

$$\rho_{x,y} = \frac{\sigma_{xy}}{\sigma_x \sigma_y} \tag{14.11}$$

Table 14.6 Matrix of Daily Correlation Between Temperature and Contaminants

	Temperature Correlation Factor				
Day	**O_3**	**SO_2**	**NO_2**	**CO**	**PM_{10}**
1	0.80970726	−0.27597056	−0.37426877	−0.17383756	0.07744498
2	0.55564806	−0.06518836	−0.23827481	−0.20677151	0.04262367
3	0.83500249	−0.06021755	0.30902205	−0.08869784	0.67519982
4	0.63555931	0.37361975	0.2043938	0.22048715	0.49305813
5	0.8320736	0.1258941	−0.35229213	−0.26758878	0.38251252
6	0.84619483	−0.0765304	−0.29050285	−0.31843396	0.32656526
7	0.75141564	0.34730821	0.33906312	0.11575915	0.57578389
8	0.65109367	0.22330793	−0.05182222	−0.17205638	0.14659114
9	0.55294346	0.28079038	0.70977252	0.30061309	0.52635791
10	0.73807656	−0.06647247	0.14089243	−0.07253488	0.44263155
11	0.61509103	0.02734897	0.41737093	0.03498226	0.41270907

Table 14.7 Correlation Coefficients Between Temperature and Contaminants

Total Correlation Factor	
O_3	0.64521348
SO_3	0.03116359
NO_2	0.08054113
CO	0.1236647
PM_{10}	0.18371268

- Where σ_{xy} is the variance of the two variables to be evaluated.
- $\sigma_x\sigma_y$ is the product of the deviation of each of the variables.

The correlation between temperature and contaminants in the entire database is shown in Table 14.5. The total correlation factor is shown in Table 14.7.

Table 14.6 expresses the correlation matrix between PM_{10} and the other pollutants per day, as calculated using Eq. (14.11).

14.4.2 FUZZY C-MEANS

The fuzzy C-means algorithm was implemented in Matlab with three scenarios, which had different numbers of centroids that are $V = 3.4, 7$. In Fig. 14.12 it can be observed clearly they are grouped into three classes, where the first class consists of high concentrations of PM_{10}, a second class was formed with average concentrations of PM_{10} and high concentrations of SO_2, and the third class with low concentrations of PM_{10} and an average concentration of SO_2.

In Fig. 14.13 it can be observed that cluster 1 and cluster 2 of the previous graph were divided and a fourth class was created, constituted of the most dispersed data of PM_{10} and O_3.

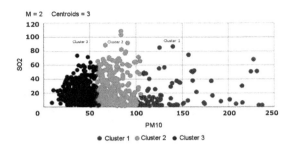

FIG. 14.12

Grouping of concentrations.

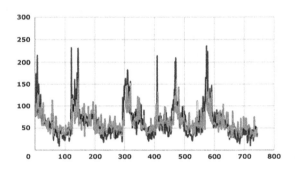

FIG. 14.13

Prediction results.

14.4.3 PREDICTION THROUGH A STATISTICAL DISTRIBUTION OF GAUSSIAN PROCESS

This section uses Weka for prediction and uses the Gaussian process algorithm for regression (GPR). This is a nonlinear Bayesian tool for estimating and detecting, which provides a high degree of confidence in its estimates and predictions [9]. GPR is a Bayesian tool for learning machines to predict the posterior probability for output (b *) given by input (X *) and a training set, which is represented in the equation

$$D = \{x_i, b_i\}_{i=1^n}, con\_x \in \left|R^d yb \in \right|R \tag{14.12}$$

In GPR a function with real values (latent function) is assumed. This is generated by a Gaussian process. The function is also Gaussian, mean zero, and its covariance is given by the function $K(x, x')$. The covariance function, which is also called the kernel, relates each point of entry and the characteristics that describe the Gaussian process.

Once the exit points $b = [b_1, b_2, \ldots b_n]^T$ are labeled (x*), the Bayes theorem is used to calculate the posterior probability density of the latent function [9]:

$$p(f, f(x^*) \mid D, x^*) = \frac{p(b \mid f, X) p(f, f(x^*) \mid X, x^*)}{p(b \mid X)} \tag{14.13}$$

Table 14.8 Percentage of Error and Correlation

Net	Centroids	MAE	MSE	RMS	Correlation
MLP	4	0.1359	0.1691	0.4112177	0.775
GPR	3	0.1504	0.18777	0.43332436	0.7619
GPR	7	0.1767	0.2368	0.486621	0.7767
GPR	4	0.1295	0.1591	0.39887341	0.8504

The Weka application could be used to investigate the behavior of the cluster in a classifier, specifically the Gaussian process, in order to predict its evolution; the K-Folks was used to generate the training and test vectors (see Table 14.8).

The prediction of PM10 of this one can be seen in Fig. 14.13, where the dark line is the real value and the light line is the value that was predicted.

14.5 **CONCLUSIONS**

A model of optimization of water savings was proposed, through a network of wireless sensors that allowed the consumption of water to be conserved from the optimization of irrigation times and waiting times between each irrigation, using metaheuristics and considering variables such as humidity, ambient temperature, and the humidity of the roots of the plants. The metaheuristic algorithm functioned by looking for the minimum value of the target function (fitness). When the fitness value was between two preset numbers (which we call the desired fitness window) then the algorithm stops and adjusts the times (watering and waiting) on the solenoid valves to keep them powered. The system continually takes samples of the moisture in the roots to match it with the moisture curve calculated from the times obtained by the algorithm. If there is a considerable difference, then fitness has a value outside the desired fitness window, so the algorithm will recalculate time values, considering that external factors might have changed, such as temperature and the humidity environment, that affected the moisture behavior in the soil.

The results from the air quality monitoring network that was used for the proposed system were: first, partitioning the region of interest into several smaller areas to improve the management of the large amount of data collected from the system, and an best coordination of the different components involved; the cluster head was then implemented in each area, which formed clusters with the nodes in their respective areas, collected their data, aggregated and sent this data to the base station; then the sensor nodes were scattered randomly in the different areas, which obtained the data and sent them to the header or head of the grouping; multiple base stations were also used to collect data from cluster heads and group them together to transmit through the gateway; and finally, the gateway collected the results from the base stations and transmitted them to the central database. This system used an air quality index to categorize the different levels of contamination. Also associated significant and intuitive colors were used for the different categories; therefore, the state of air pollution can be communicated to the user very easily. The system also uses an AQI to evaluate the level of health concern for a specific area. The wireless air pollution monitoring system used a novel technique for data aggregation to address the challenge of minimizing energy consumption in the WSN. Quartiles were also used to summarize a list of readings of any length to three values, which reduces the amount of data to be transmitted to the base station, thus reducing the required transmission power and, at the same time, representing the original values accurately.

REFERENCES

[1] IEEE Standard. Wireless medium access control (MAC) and physical layer (PHY). New York, NY: IEEE; 2006. https://standards.ieee.org/findstds/standard/802.15.4-2006.html.

[2] Dargie W, Poellabauer C. Fundamentals of wireless sensor networks: theory and practice. Wiley; 2010. ISBN: 978-0-470-99765-9.

[3] Karl H, Willig A. Protocols and architectures for wireless sensor networks. New York: Wiley; 2005. ISBN: 978-0-470-09510-2. 526 p.

[4] Sohraby K, Minoli D, Znati T. Wireless sensor networks: technology, protocols, and applications. New Jersey: John Wiley & Sons; 2006. ISBN: 9780471743002. https://doi.org/10.1002/047011276X.

[5] Xbee. Xbee-Pro, ZB R.F. modules, Minnetonka: Digi International; 2012. https://www.digi.com/.

[6] Mishra SA, Tijare DS, Asutkar M. Design of energy aware air pollution monitoring system using WSN. Int J Adv Eng Technol 2011;1:107–16.

[7] Bielsa A. http://www.libelium.com; 2011. December 27, Waspmote technical guide, Document version v7.5, Libelium Comunicaciones Distribuidas S.L. http://www.libelium.com/smart_city_air_quality_urban_traffic_waspmote/.

[8] Yerra R, Sami Baig M, Mishra R, Pachamuthu R, Desai U, Merchant S. Real time wireless air pollution monitoring system. Next Gen Wirel Netw Appl 2011;2:370–5. https://doi.org/10.21917/ijct.2011.0051.

[9] Jamil MS, Jamil M, Mazhar A, Ikram A, Ahmed A, Munawar U. Smart environment monitoring system by employing wireless sensor networks on vehicles for pollution free smart cities. Process Eng 2015;107:480–4. https://doi.org/10.1016/j.proeng.2015.06.106.

FURTHER READING

[1] Al-Karia JN, Kamal AE. Routing techniques in wireless sensor networks: a survey. IEEE Wirel Commun 2012;11(6):6–28.

[2] Augusto JC, Callaghan V, Cook D, Kameas A, Satoh I. Intelligent environments: a manifesto. Human-Centric Computing and Information Sciences 2013;3:12. https://doi.org/2192-1962-3-12. Springer, Berlin, Heidelberg.

[3] Baquero RS, Rodríguez JG, Mendoza S, Decouchant D. in: Towards a uniform sensor-handling scheme for ambient intelligence systems. 2011 8th International conference on electrical engineering computing science and automatic control (CCE), 26–28 October 2011, Merida City, Mexico. IEEE; 2011.

[4] Barcenas FEP. Sistema de Monitoreo Móvil de la Calidad del Aire Utilizando Redes Vehiculares de Transporte. Querétaro, Mexico: Universidad Autónoma de Querétaro; 2013.

[5] Cortina-Januchs MG, Quintanilla-Domínguez J, Andina D, Vega-Corona A. in: Prediction of PM10 concentrations using fuzzy C-means and ANN. World Automation Congress (WAC); 2012. p. 1–6.

[6] Das S. Pattern recognition using the Fuzzy c-means technique. Int J Energy Inf Commun 2013;4:1–14.

[7] El-Ghazali T. Metaheuristics from design to implementation. New Jersey: John Wiley and Sons, University of Lille CNRS INRIA; 2009.

[8] Omer Farooq M, Kunz T. Operating systems for wireless sensor networks: a survey. Sensors (Basel) 2011;11(6):5900–30.

[9] Fu C, Jiang Z, Wei W, Wei A. An energy balanced algorithm of LEACH protocol in WSN. Int J Comput Sci Issues 2013;10(1):354–9.

[10] García-Hernando A-B, Martínez-Ortega J-F, López-Navarro J-M, Prayati A, Redondo-López L, editors. Problem solving for wireless sensor networks computer communications and networks. España: Springer; 2008.

[11] Ibarra-Berastegi G. From diagnosis to prognosis for forecasting air pollution using neural networks: air pollution monitoring in Bilbao. Environ Model Softw 2008;23:622–37.

[12] Kukkonen J. Extensive evaluation of neural network models for the prediction of NO_2 and PM10 concentrations, compared with a deterministic modelling system and measurements in central Helsinki. Atmos Environ 2003;37:4539–50.

[13] Sohraby K, Minoli D, Znati T. Wireless sensor networks technology, protocols, and applications. New York: Wiley; 2007.

[14] Khedo KK, Perseedoss R, Mungur A. A wireless sensor network air pollution monitoring system. Int J Wirel Mobile Netw 2010;2(2):31–45.

[15] Kim KT, Lyu CH, Moon SS, Youn HY. Tree-based clustering (TBC) for energy efficient wireless sensor networks. Advanced information networking and applications workshops (WAINA); 2010. p. 680–5.

[16] Yadav L, Sunitaha C. Low energy clustering hierarchy in wireless sensor network (LEACH). Int J Comput Sci Inf Technol 2014;5(3):4661–4.

[17] Li J, Andrew LL, Foh CH, Zukerman M, Chen HH. Connectivity, coverage and placement in wireless sensor network. Sensors (Basel) 2009;9(10):7664–93.

[18] Mujawar TH, Bachuwar VD, Suryavanshi SS. Air pollution monitoring system in solapur city using wireless sensor network, proceedings published by. Int J Comput Appl 2013;11–5.

[19] Pearce F. El futuro TUNZA, la revista del PNUMA para los jóvenes. Tomo 5, 7, La revista del PNUMA para los jóvenes 2010.

[20] Prasad RV, Baig MZ, Mishra RK, Rajalakshmi P, Desai UB, Merchant SN. Real time wireless air pollution monitoring system. ICTACT J Commun Technol 2011;2(2):370–5.

[21] Rabiner-Heinzelman W, Chandrakasan A, Balakrishnan H. Energy-efficient communication protocol for wireless microsensor networks. In: Proceedings of the 33rd annual Hawaii international conference on system sciences. Vol. 2. 2000. p. 10. https://doi.org/10.1109/HICSS.2000.926982. 2000.

[22] Rehman O, Javaid N, Manzoor B, Hafeez A, Iqbal A, Ishfaq M. Energy consumption rate based stable election protocol (ECRSEP) for WSNs. International workshop on body area sensor network (BASNet-2013), Procedia Computer Science. 19; 2013. p. 932–7.

[23] Shaughnessy SAO, Evett SR. Canopy temperature based system effectively schedules and controls center pivot irrigation of cotton. Agric Water Manag 2010;97(9):1310–6.

[24] Sivanandam SN, Deepa SN. Introduction to genetic algorithms. Berlin, Heidelberg: Springer-Verlag; 2008.

[25] Smaragdakis G, Ibrahim M, Bestavros A. SEP: a stable election protocol for clustered heterogeneous wireless sensor networks. Second international workshop on sensor and actor network protocols and applications. vol. 3. Massachusetts: Boston; 2004.

[26] Kyung T-K, Chang H-L, Sung S-M, Hee Y-Y. Tree-based clustering (TBC) for energy efficient wireless sensor networks. IEEE 24th international conference on advanced information networking and applications workshops. Australia: Perth; 2010. ISBN: 978-0-7695-4019-1.

[27] Wang X, Yang W, Wheaton A, Cooley N, Moran B. Efficient registration of optical and IR images for automatic plant water stress assessment. Comput Electron Agric 2010;74(2):230–7.

[28] Yaacoub E, Kadri A, Mushtaha M, Abu-Dayya A. Air quality monitoring and analysis in qatar using a wireless sensor network deployment. Wireless communications and mobile computing conference, Cagliari, Sardinia, Italy, IEEE; 2013. p. 1–6.

[29] Bani Yassein M, Al-zou'bi A, Khamayseh Y, Mardini W. Improvement on LEACH protocol of wireless sensor network (VLEACH). Int J Digit Content Technol Appl 2009;3(2):132–6.

Index

Note: Page numbers followed by *f* indicate figures, and *t* indicate tables.